P9-BJT-243

DATE DUE

FEB 15 1990			
APR 25 1995			
GAYLORD			PRINTED IN U.S.A.

TABLE OF CONTENTS

CREDITS AND ACKNOWLEDGEMENTS

Kértesz-Sáringer, M.; E. Mészáros; and T. Várkonyi, "Technical Note on the Size Distribution of Benzo(a)Pyrene Containing Particles in Urban Air," *Atmospheric Environment*, 1971, 5:429-431.

Kneip, Theo. J.; Merril Eisenbud; Clifford D. Strehlow; and Peter C. Freudenthal, "Airborne Particulates in New York City," *Air Pollution Control Association Journal*, 1970, 20:144-149.

Lippmann, Morton; and Agis Kydonieus, "A Multi-Stage Sampler for Extended Sampling Intervals," *American Industrial Hygiene Association Journal*, 1970, 31:730-737.

Lundgren, Dale A., "Atmospheric Aerosol Composition and Concentration as a Function of Particle Size and of Time," *Air Pollution Control Association Journal*, 1970, 20:603-608.

Lundgren, Dale A., "Determination of Particulate Composition, Concentration and Size Distribution Changes with Time," *Atmospheric Environment*, 1971, 5:645-651.

Mammarella, L., "Evaluation of Aerosol Pollution Using a Three Device System," *Pure and Applied Chemistry*, 1970, 24:749-755.

Mammarella, L., "The Granulometric Classification of Aerosols Using a New Type of Cascade Impactor (Four Stage Aerosol Differential Sampler)," *Pure and Applied Chemistry*, 1970, 24:715-719.

Moss, Owen R., "Shape Factors for Airborne Particles," *American Industrial Hygiene Association Journal*, 1971, 32:221-229.

Moroz, W.J.; V.D. Withstandley; and G.W. Anderson, "A Portable Counter and Size Analyzer for Airborne Dust," *The Review of Scientific Instruments*, 1970, 41:978-983.

Morse, K.M.; H.E. Bumsted; and W.C. Janes, "The Validity of Gravimetric Measurements of Respirable Coal Mine Dust," *American Industrial Hygiene Association Journal*, 1971, 32:104-114.

Sehmel, G.A., "Particle Sampling Bias Introduced by Anisokinetic Sampling and Deposition within the Sampling Line," *American Industrial Hygiene Association Journal*, 1970, 31:758-771.

PREFACE

In the United States air pollution reaches toxic thresholds in almost all urban areas. Its cost in terms of life and health is extreme. According to various governmental studies, the annual cost of damage resulting from polluted air is over $13 billion. If, for example, pollution could be cut by 50 percent, newborn babies would have an additional life expectancy of three to five years and deaths would be reduced 4.5 percent. Air pollution is also a factor in the development of diseases ranging from respiratory disorders to cirrhosis of the liver, and polluted air intensifies allergic reactions. Serious long-term research and conscientious political action are needed if these ominous trends are to be reversed.

Volume VII in MSS' continuing series on air pollution presents current research on the characteristics and detection of airborne particulate pollutants. Both the shape and size of particulates are discussed, and the importance of the time factor in determining particle concentration is pointed out. Analytic methods for counting and classifying various types of particulate pollutants are presented in detail.

Atmospheric Aerosol Composition and Concentration as a Function of Particle Size and of Time

Dale A. Lundgren

Diurnal variations in the size distribution, composition, and concentration of atmospheric aerosols are of relatively unknown magnitude, largely because of the difficulty in making such measurements. Most atmospheric particulate data is that based on 24-hr high-volume sampler data, such as that obtained by the National Air Sampling Network. This data is of great historic value and very necessary for determination of long-range trends, seasonal variations, and overall concentration and composition averages. But, the real effect of particulate pollutants on human health, vegetation, materials, or visibility cannot be adequately determined by high-volume filter data alone; nor can a basic understanding of aerosol formation be gained from measurement averages obtained over periods as long as the basic aerosol formation process itself.

The entire atmospheric pollutant cleansing process is dependent upon the formation, growth, and removal of particulate matter. Photochemical aerosol formation is an outstanding example of atmospheric reaction and dissipation of gaseous pollutants to form particulate pollutants.

This study has been an effort to measure aerosol composition, concentration, and particle size distribution changes taking place over reasonably short time periods in order to determine and better understand phenomena such as photochemical aerosol formation. This paper describes the sampling-analysis method developed, presents data which was obtained, relates the obtained particulate measurements with gaseous pollutant and meteorological measurements simultaneously taken, and reaches several conclusions.

All data reported in this paper was obtained at the Riverside Campus of the University of California—about 50 miles east of downtown Los Angeles.

Experimental Procedure

During the first half of November 1968, samples of atmospheric particulate matter were obtained using a Lundgren Impactor[1] and several filter samplers. Impactor samples were used to determine the size distribution of the total particulate plus various particulate components, namely: sulfate, nitrate, lead, and iron. Filter samples were used to determine diurnal variations in total particulate and particle sulfate, nitrate, and lead.

A Royco Instruments Model 230 photometer was used to continuously monitor the total particulate light-scattering. Continuous measurements were also made of wind speed, wind direction, dry bulb temperature, wet bulb temperature, and six gaseous pollutants—carbon monoxide, nitrogen dioxide, nitric oxide, hydrocarbon, oxidant, and peroxyacetyl nitrate.

Size fractionated aerosol samples, amenable to both physical (including microscopic) and chemical analysis, were obtained using a special sampling procedure which was developed for the Lundgren impactor.

Two-inch diameter filters fitted with glass fiber filter paper, were used to obtain total particulate samples for weight determinations and analysis of particulate sulfate, nitrate, and lead. Particulate iron was also analyzed for but the data was not used because iron background of the filter media was too high and variable. Samples were run every 4 hours starting at 12, 4, and 8

A.M. and P.M. A high-volume sampler was also run each day on a 24-hr basis. The high-volume sampler was operated and analyzed in the manner recommended by Jutze and Foster in the TR-2 Air Pollution Measurements Committee Report.[2]

All filter and impactor samples were extracted for analysis by the methods detailed in the Air Pollution Measurements Committee report. Both sulfate and nitrate were analyzed as described therein. Iron and lead were also extracted as described but were run by atomic absorption spectrophotometry.

The 24-hr high-volume sampler filter was analyzed and used as a check against the six corresponding 4-hr low-volume sampler filters. The impactor samples were checked in a similar manner against low-volume sampler filters run side by side over the same time period and at the same flow rate. Because of the setup time required for the impactor, samples were run only over 16-hr periods, from 4 P.M. to 8 A.M.

All the gas pollutant and meteorological measurements made are well described in the literature and will not be described further here. These measurements were all continuously recorded and then averaged over 4-hr periods which correspond to the particulate sampling period.

Impactor Results

Ten size-fractionated samples of atmospheric particulate matter were obtained using the Lundgren impactor. These samples were analyzed as follows:

1. Particulate dry weight distribution determined (based on the particle weight obtained after desiccating the samples for 6 hr at room temperature and pressure in a desiccator containing drierite, or $CaSO_4$).
2. Water-soluble fraction extracted and analyzed for sulfate and nitrate by standard wet chemical methods.[2]
3. Nitric acid fraction extracted and analyzed (in addition to the water-

Table I. Tabulation of impactor analysis results.

Run	Total particulate µg/m³	MMD(µ)	σg	Sulfate µg/m³	MMD(µ)	σg	Nitrate µg/m³	MMD(µ)	σg	Lead µg/m³	MMD(µ)	σg	Iron µg/m³	MMD(µ)	σg
1	92	1.0	7	13.1	0.6	4	18.1	1.1	3	0.80	0.6	5	0.53	2.2	10
2	64	0.5	8	9.1	0.3	6	12.8	0.6	3	0.52	0.9	6	0.45	1.1	12
3	40	0.7	16	4.5	~0.15	—	3.7	0.9	3	0.48	0.5	7	0.61	0.6	20
4	144	0.5	7	12.6	~0.15	7	18.0	1.0	4	0.85	0.5	5	0.89	1.1	10
5	141	0.6	10	12.6	~0.15	—	29.9	0.7	2	0.79	0.4	6	1.28	1.7	7
6	69	4.5	10	7.3	~0.1	8	6.2	0.9	3	0.35	0.6	20	0.99	2.2	11
7	79	1.0	8	6.0	~0.15	5	8.6	0.7	3	0.80	~0.2	10	0.49	5.6	3
8	63	1.2	9	10.5	0.5	3	6.4	1.2	3	0.52	0.6	6	0.52	1.7	8
9	75	2.1	15	2.9	0.6	4	7.7	0.7	4	0.47	0.3	15	0.66	9.3	5
10	47	1.7	15	4.6	~0.2	6	5.3	1.8	4	0.30	0.9	20	0.46	1.4	10
Data avg. (Fig. 2)	82	0.9	11	8.3	~0.3	4	11.7	0.8	3	0.59	0.5	7	0.69	2.2	8
Data range	40–144	0.5–4.5	7–16	2.9–13.1	~0.1–0.6	3–8	3.7–29.9	0.6–1.8	2–4	0.30–0.85	~0.2–0.9	5–20	0.45–1.28	0.6–9.3	3–20
Summary of other impactor analysis results obtained during 1968															
Data range	25–318	0.7–4.0								0.40–2.84	~0.1–1.1				

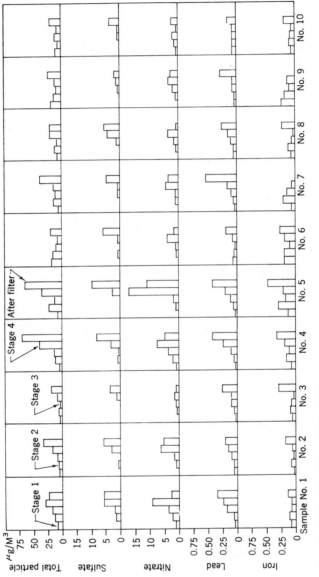

Figure 1. Plot of impactor analysis results.

12

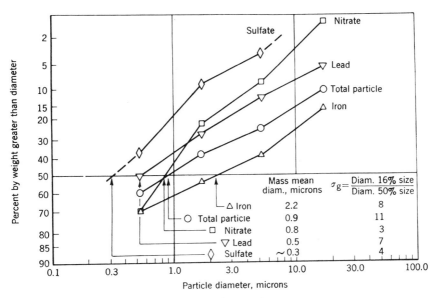

Figure 2. Average size distributions for 10 impactor samples.

soluble fraction) by atomic absorption spectrophotometry for lead and iron.

Results for these ten tests are plotted in Figure 1 as μg of the particulate, nitrate, etc., per m³ of air sampled—for each of the four impactor stages plus the impactor after filter. Table I summarizes the particulate size distribution data obtained from the analysis results. At the end of Table I, other impactor analysis data obtained during 1968 is also summarized. In all cases, samples were run from 4 P.M. of one day until 8 A.M. of the next. Air sampling rate was held constant at 2.9 cfm for all runs except number one which had a flow of 4.0 cfm. Based on a density one spherical particle and a 2.9 cfm flow rate, the impactor classification diameters are: greater than 17μ on stage one, 5.2–17μ on stage two, 1.7–5.2μ on stage three, 0.5–1.7μ on stage four, and less than 0.5μ on the after filter. None of these test runs were

divided into time fractions; therefore, the numbers given represent averages over the 16-hr sampling periods.

The data for these ten runs were used to obtain an average distribution for total particulate weight, sulfate, nitrate, lead, and iron. Figure 2 is a log probability plot of these weight averages vs. diameter. Again, the diameter is based on a density one spherical particle. Microscopic examination of the collected particulates indicates this assumption to be reasonably good for photochemical aerosols around one micron diameter. In noting the differences in the mass median diameters and distributions shown in Figures 1 and 2, it is important to remember that they are all based on analysis of the same samples, and not different samples taken at different times or locations.

Data shown in Figure 1 and Table I are not highly accurate. Total particulate weight losses in the impactor averaged 10% and ranged from 0 to

13

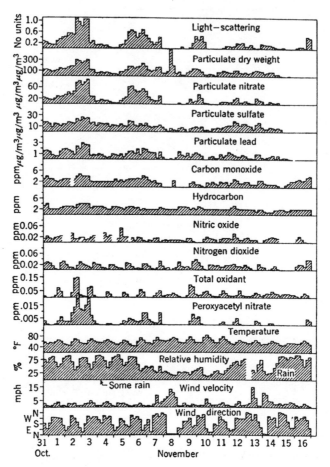

Figure 3. Particulate, gas, and meteorological measurement averaged over 4-hr periods.

20% for the ten tests. Average sulfate loss was 25%, average nitrate loss was 32%, and average lead loss was 36%. All losses were computed by comparing the sum of the impactor stages plus after filter with a total filter sample. Average iron loss could not be determined because of high filter iron background. The individual impactor after filter results of Figure 1 may have an error equivalent to about ±0.1 μg/m³ of iron because of the filter background variability.

Although chemical analysis of the impactor stage plus after filter recovered (on the average) only about 70% of the particulate component of interest, the size distribution and mass mean diameter data of Figure 2 is considered reasonably accurate, quite probably within ±20% for the mean diameters. The relationship of the various curves to each other should be quite accurate because analysis losses were similar and appeared to be the same percentage for all stages.

14

Table II. Tabulation of correlation coefficients.

	Light-scattering	Particle weight	Particle weight (data omitted)	Nitrate	Sulfate	Lead	PAN	Oxidant	Hydrocarbon	CO	NO	NO$_2$	Relative humidity	Wind velocity
Particle light-scattering	—	0.71	0.94	0.94	0.71	0.82	0.89	0.30	0.63	0.69	0.04	0.36	0.22	-0.26
Particle dry weight	0.71	—	—	0.69	0.52	0.57	0.66	0.29	0.42	0.48	-0.05	0.19	0.00	0.15
Particle dry weight (data omitted)	0.94	—	—	0.92	0.72	0.80	0.86	0.36	0.55	0.72	-0.02	0.35	0.11	-0.13
Particle nitrate	0.94	0.69	0.92	—	0.74	0.80	0.81	0.26	0.60	0.74	0.05	0.26	0.30	-0.28
Particle sulfate	0.72	0.52	0.72	0.74	—	0.71	0.60	0.09	0.63	0.73	0.26	0.38	0.48	-0.33
Particle lead	0.82	0.57	0.80	0.80	0.71	—	0.69	0.02	0.69	0.76	0.03	0.52	0.36	-0.45
Peroxyacetyl nitrate	0.89	0.66	0.86	0.81	0.60	0.69	—	0.57	0.53	0.53	-0.09	0.28	0.05	-0.19
Total oxidant	0.30	0.29	0.35	0.26	0.09	0.02	0.57	—	-0.05	0.02	0.01	-0.17	-0.39	0.12
Hydrocarbon	0.63	0.42	0.55	0.59	0.63	0.69	0.53	-0.05	—	0.58	0.14	0.44	0.36	-0.37
Carbon monoxide	0.69	0.48	0.72	0.74	0.73	0.76	0.53	0.02	0.58	—	0.13	0.43	0.42	-0.38
Nitric oxide	-0.04	-0.05	-0.02	0.05	0.26	0.03	-0.09	0.01	0.14	0.13	—	-0.23	0.20	-0.02
Nitrogen dioxide	0.36	0.19	0.35	0.26	0.38	0.52	0.28	-0.17	0.44	0.43	-0.23	—	0.20	-0.46
Relative humidity	0.22	-0.00	0.11	0.30	0.48	0.35	0.05	-0.40	0.36	0.42	0.20	0.20	—	-0.43
Wind velocity	-0.26	0.15	-0.13	-0.28	-0.33	-0.45	-0.19	0.12	-0.37	-0.39	-0.02	-0.47	-0.43	—

Filter Results

During the 15-day period from about Nov. 1 to Nov. 15, 1968, samples of atmospheric particulate matter were obtained over 4-hr periods (a total of 90 separate samples). Each particulate sample was analyzed for particulate dry weight, nitrate, and sulfate in the water-soluble portion and for total particulate lead content. Iron was also determined but was not included because of a relatively high and variable iron background in the glass fiber filters used. In addition to, and for comparison with, the particulate analysis results, total particle light-scattering, wind velocity, wind direction, air temperature, relative humidity, and the concentration of six gaseous pollutants, were obtained, averaged over the 4-hr particle sampling periods, and all plotted in Figure 3. The six gaseous pollutants plotted—carbon monoxide, nitrogen dioxide, nitric oxide, hydrocarbon, oxidant, and peroxyacetyl nitrate—are those routinely measured by the air monitoring station at the Statewide Air Pollution Research Center of the University of California—Riverside.

Methods of filter analysis are all standard to the air pollution field and were discussed in the Experimental Procedure section. All plotted particulate results were based on analysis of 2-in glass fiber filters through which a volume of about 35 m³ of air had been sampled over the 4-hr periods. Based on a 100 μg/m³ dust level, this corresponded to 3.5 mg of collected particulate. Using a semimicro balance, weight reproducibilities of ±0.05 were obtained (this was also true for the impactor weight reproducibilities). At this dust level, particulate analysis reproducibility tests produced differences averaging ±5%, suggesting an overall analysis error of perhaps ±10%. At dust levels below 20 μg/m³ the analysis error is probably great but was not actually determined.

During the November sampling sequence, only the days of Nov. 2, 3, 6, and 8 were above the yearly suspended dust average of 150 μg/m³. Nov. 8 had a high average level because a very high dust loading (498 μg/m³) resulted during a very windy 4-hr period in which dry ground sand and soil were aerosolized. The visibility was fairly high and the gaseous pollutant levels were fairly low for this windy period, as is usually the case for high desert wind conditions.

Suspended particulate averaged about 100 μg/m³ which is normal for that period of the year. The days of Nov. 2, 3, and 6 are more comparable to summer days where the suspended particulate average is about 175.

The very clean day observed on Nov. 15 is normally found only in the winter following a heavy rain. Days this clean occur about 1% of the time in Riverside. Days as dirty as Nov. 2 and 3 would occur about half of the time in summer and may represent about one-third of the days in a year.

Riverside has about the same average suspended dust level as downtown Los Angeles but with greater variation.

Data Correlation Results

A general correlation is seen to exist between several of the measured parameters plotted in Figure 3. A quantitative estimate of the relationship between these measured parameters was determined by assuming a linear relationship, determining the best fitting straight line (by the method of least squares) and calculating a correlation coefficient R. These R values, listed in Table II, ranged from near zero to a high of 0.94. Although an R value of 0.94 for 90 data points is very good, it does not necessarily allow a satisfactory estimate of one pollutant value from the other measured value.

A visualization of the relationships between various pollutant measurements can best be obtained by plotting the values. In Figure 4, particulate nitrate is plotted against total light-scattering.

16

The correlation between particulate dry weight and light-scattering illustrates how much a relationship can be affected by two points which are of extreme or very unusual nature. The two points in question are for Nov. 8, where dust loadings of 498 and 198 were obtained during a relatively clear but extremely windy afternoon. The Table II column titled "Particle dry weight (Data omitted)" is the case with these two points omitted. Throwing out these two points increased the R value from 0.71 to 0.94 and greatly changed the 95% confidence interval. Considering the effect two data points can have on a correlation coefficient, the quantitative R values of Table II should only be qualitatively compared. Light-scattering does correlate well with particulate nitrate, peroxyacetyl nitrate (PAN), and total particulate if the windy days are thrown out. There is also a low but meaningful correlation with hydrocarbons and carbon dioxide. Particulate dry weight naturally correlates with particulate components but also shows a high correlation with PAN and a low but meaningful correlation with hydrocarbon and carbon dioxide. In general, the same correlations exist for all particulate measurements. PAN correlates quite well with particle nitrate and light-scattering, and fairly well with sulfate, lead, total oxidant, hydrocarbon, and carbon dioxide. An interesting observation is that total oxidant correlates fairly well only with PAN. Hydrocarbon and carbon monoxide are very similar in that both correlate about equally well with other pollutants, mainly the particle measurements, PAN, and somewhat with nitrogen dioxide— but not with nitric oxide. Nitric oxide, in fact, does not correlate well with anything measured. Nitrogen dioxide correlates best with lead and somewhat with hydrocarbon and carbon dioxide. Relative humidity had a low positive correlation with sulfate and a low negative correlation with total oxidant and wind velocity. Wind velocity had a

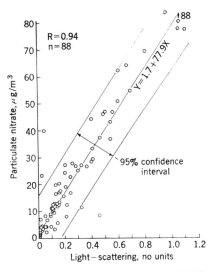

Figure 4. Relationship between particulate nitrate content and total aerosol light-scattering.

slight negative correlation with everything, meaning windy days tended to be clean days except for the blowing sand.

Measurement accuracy for the gaseous pollutant data used is not known and will not be discussed. Probably the nitric oxides are the least accurately measured and the poor correlation obtained is partially because of this.

Summary and Conclusions

Impactor samples of Riverside air were obtained during times of heavy smog and times of clear air. Microscopic examination indicated a great difference in the number and type of particulates present, especially in the 1μ size range. On days of heavy smog, very hygroscopic, crystalline-like particles were found to comprise over half the particulate dry weight in the $0.5-1.5\mu$ diameter size range. These crystalline particles were analyzed by X-ray diffraction and identified as ammonium nitrate. On days when the wind shifted and heavy smog quickly

Table III. Tabulation of 24-hour suspended particulate analysis results.

Date (1968)	Analysis results—$\mu g/m^3$				Results as % of total particulate		
	Total Particulate (dry wt.)	Water-soluble	Sulfate	Nitrate	Water-soluble,	Sulfate	Nitrate
8–1	287	135	31.9	51.0	47.0	11.1	17.8
2	257	122	32.8	42.0	47.5	12.8	16.3
3	244	125	31.9	47.1	51.2	13.1	19.3
4	193	84	25.3	31.8	43.5	13.1	16.5
5	212	83	21.4	30.5	39.2	10.1	14.4
6	273	126	31.4	53.6	46.2	11.5	19.6
7	201	75	14.4	36.8	37.3	07.2	18.3
8	164	64	11.0	33.4	39.0	06.7	20.4
9	158	60	10.0	30.2	38.0	06.3	19.1
10	195	85	19.3	30.2	43.4	09.8	15.4
12	140	37	12.8	11.0	26.4	09.1	07.9
13	97	25	7.3	8.8	25.8	07.5	09.1
14	189	45	14.2	16.3	23.8	07.5	08.6
15	218	81	20.8	37.1	37.2	09.5	17.0
16	108	30	11.7	13.4	27.8	10.8	12.4
19	146	61	16.0	28.3	41.8	11.0	19.4
20	74	17	3.6	7.2	23.0	04.9	09.7
21	90	24	2.2	7.5	26.7	02.4	08.3
22	137	55	4.1	11.8	40.1	03.0	08.6
23	84	20	1.7	4.0	23.8	02.0	04.8
24	124	38	4.3	5.7	30.6	03.5	04.6
26	149	40	9.1	9.0	26.8	06.1	06.0
27	168	54	7.4	20.7	32.1	04.4	12.3
28	288	94	18.4	28.5	32.6	06.4	09.9
29	207	72	14.6	23.7	34.9	07.1	11.4
Data avg.	176	66	15.1	24.8	37.5	8.6	14.1
Data range	74–288	17–135	1.7–32.8	4.0–53.6	23%–51%	2%–13%	5%–20%

No data for August 11, 17, 18, 25, 30, and 31.

moved in, dramatic changes in the particulate appearance were observed at the same time that increases in PAN and oxidant were recorded. Chemical analysis of these impactor samples showed the water-soluble fraction to be very high in nitrate and fairly high in sulfate.

The November data of Figure 3 contains only a few days of heavy smog. Therefore, the data of Table III is included to better illustrate the high percentage of water-soluble particulate, especially nitrate, in the Riverside, Calif. aerosol. These results are for standard, 24-hr high-volume filter samples obtained during Aug. 1968 and analyzed as described previously. In general, days with high oxidant and PAN had high total particulate with a water-soluble fraction of 40–50%. Days of lowest oxidant and PAN had water-soluble fractions (based on particle dry weight) of only 25%. Diurnal variations made it difficult to quantitate any relationship between particulate, gas, and meteorological measurements on this 24-hr average basis.

The results of this study are the basis for the following comments about Riverside-type photochemical aerosol.

Photochemically (or atmospherically) produced particulate matter is mainly a water-soluble nitrate compound. It is a major factor in causing visibility reduction in the inland Los Angeles air basin. This photochemical aerosol is of a very hygroscopic nature and often has a mass mean diameter of about 1μ. The hygroscopic nature and size (about the wavelength of light) of photochemical aerosol particles enable them to acquire and hold fairly large quantities of water and greatly affect visibility. Humidity has been shown to have a great effect on the light scattered by an aerosol.[3] This factor is important in explaining the heavy haze which develops in the Riverside-San Bernardino area, especially when moist ocean air is mixed with the basin photochemical aerosol. This haze will persist even after the air temperature has risen enough to decrease the relative humidity well below 100%.

A particulate lead mass mean diameter of about 1μ has been measured on several summer days of high smog level. This indicates that the atmospheric aerosol being sampled in Riverside must have an age (or atmospheric residence time) of about one or more days to allow the 0.1μ lead-containing auto exhaust particles to grow by agglomeration with other particles to the mean size of 1μ.

The very significant difference in the particle size distribution slope for lead and nitrate particulate fractions shows that the nitrate aerosol does not grow by pure physical agglomeration as did the lead aerosol. The similarities in the nitrate and sulfate size distribution slopes and their greater size uniformity indicate both are formed in the atmosphere by a chemical condensation process rather than a physical agglomeration process.

On a summer day of high smog and west winds, the particulate matter sampled in Riverside, 50 mi east of downtown Los Angeles, may have been in the atmosphere 12–24 hr longer than the particulate sampled in downtown Los Angeles. A most obvious difference

in the particulate analysis results for these two locations is the high Riverside particulate nitrate concentration. This high nitrate is related to proper meteorological conditions, highly correlated with photochemically produced PAN, and reasonably well correlated with the concentration of gaseous pollutants from the internal combustion engine. Although a particulate nitrate of 60 $\mu g/m^3$ may seem high, the gaseous nitrate concentrations in Riverside are even higher. PAN alone reaches mass concentrations of 240 $\mu g/m^3$ (50 ppb) in Riverside on summer days of high smog level.

The data of Figures 1, 2, and 3, and the above comments, illustrate the type of atmospheric aerosol information obtainable with the described sampling methods. This, however, represents a fraction of the information that can be obtained from a more complete analysis of the impactor and filter samples. Data obtained by microscopic examination of the impactor films was subjective and was not included because of space limitations.

Once the technique was developed, obtaining good, size-classified, time-fractionated, particulate samples became the easiest part of the study.

References

1. Lundgren, D. A., "An aerosol sampler for determination of particle concentration as a function of size and time," *J. Air Poll. Control Assoc.*, 17 (4):225 (1967).
2. Jutze, G. A. and Foster, K. E., "Recommended standard method for atmospheric sampling of fine particulate matter by filter media—High-volume sampler," *J. Air Poll. Control Assoc.* 17(1):17 (1967).
3. Lundgren, D. A. and Cooper, D. W., "Effect of humidity on light-scattering methods of measuring particle concentration." *J. Air Poll. Control Assoc.*, 19(4):243 (1969).

DETERMINATION OF PARTICULATE COMPOSITION, CONCENTRATION AND SIZE DISTRIBUTION CHANGES WITH TIME*

DALE A. LUNDGREN

Abstract—A sampling method which enables determination of particulate composition and concentration from a single aerosol sample, both as a function of particle size and of time, is described. Analysis of the size-classified, time fractioned particulate samples can be accomplished by normal chemical or physical means. To illustrate this method's applicability several samples were analyzed, and results are presented for total weight distribution and particulate sulfate, nitrate, lead and iron concentration.

INTRODUCTION

THIS paper describes a method for the determination of timewise fluctuations in the size, concentration and chemical composition of particulate matter in air. Providing this data necessitates:

(1) The accretion of size fractionated samples of the aerosol particles over a desired time period.

(2) The collection of the particulate matter in such a way that it is amenable to analysis.

(3) The analysis of the particulate samples by techniques capable of providing the required information.

No single sampling instrument or method can be optimum for all possible aerosol analysis requirements. Therefore, the following sampling–analysis method was intended as a reasonable compromise for normal air pollution and industrial hygiene requirements.

SAMPLER DESCRIPTION

This sampling-analysis method is based on the use of a fairly high flow rate four stage impactor (called the Lundgren Impactor). A cross-section schematic of this sampler is shown in FIG. 1. Particles are collected by inertial impaction on the surface of the rotating collection cylinders shown. The collection drum drive can be set to provide the proper rotational speed for the desired sampling time. In all cases, a chronological collection deposit is produced with good time resolution.

The flow rate through the impactor is not fixed, and air can be sampled at any rate from about 0.5 ft^3 min^{-1}‡ (below which gravitational loss of large particles occurs) up to about 5 ft^3 min^{-1} (above which excessive pressure drop and high particle wall loss occurs). This flow rate range enables classification of particles over a wide size

---

* Presented at Symposium on Advances in Instrumentation for Air Pollution Control, sponsored by the National Air Pollution Control Administration, Department of Health, Education and Welfare, Cincinnati, Ohio, 26–28 May, 1969.

‡ 0.5 ft^3 min^{-1} = 236 cm^3 s^{-1}.

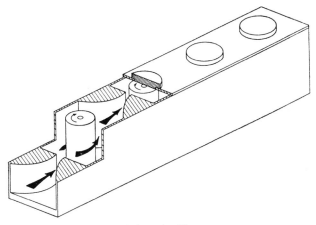

FIG. 1. Schematic of impactor.

range, as shown in FIG. 2. Collection characteristics of the impactor were determined experimentally (LUNDGREN, 1967). The impactor stage 50 per cent cut points shown in FIG. 2 were calculated from the calibration data.

SAMPLING PROCEDURE

A sampling procedure should be mated to an analysis method or methods. If several analysis methods are to be used, the sampling procedure should be versatile without being overly involved.

It was stated that particles to be collected are impacted onto the collection drum. Normally, the drum is coated with a thin film material and the particulate matter collected onto this removable drum coating material. Materials such as metal foil, sticky tapes, and various plastic films have been used. The most suitable material tried was a film of Teflon, about 0.001 in. thick. In general, the sticky tapes (such as two-sided sticky tape) were the least suitable and are not recommended.

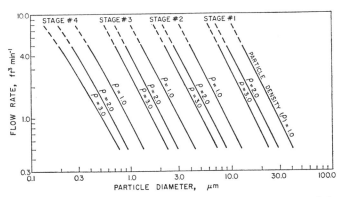

FIG. 2. Particle size classification data (based on stage 50 per cent cut-point).

21

Teflon film has the advantages of being inert, having a low chemical background, and having a low affinity for water vapor. It is very weight-stable, is not affected by most acids or bases, and is quite transparent under the microscope. Teflon films have the advantage of being easily cut into time based segments for chemical analysis.

An after filter must be used to collect particulate matter passing the fourth stage of the impactor. Desirable characteristics of an after filter include high collection efficiency, low pressure drop, good weight stability, low chemical background, and good particulate loading characteristics (filter does not plug up). A variety of filter materials have been tested to determine their overall usefulness. Based on weight stability and low chemical background, Teflon filters (sold by Millipore Corp.) are rated the best in performance. These filters can be pre-cleaned by washing in acid and water to reduce their chemical background to near zero.

If a low-cost filter medium is desired, the standard glass fiber filter medium should be used for the after filter. It has reasonable weight stability and a fairly low chemical background.

ANALYSIS PROCEDURES

Once a sample has been obtained, it is ready for analysis. Sample weights should be determined prior to microscopic viewing. Changes in particulate deposit density with time can often be seen directly on the film; these changes can be determined by particle count as a function of deposit position (or time) or by cutting the collection surface film into time based strips and analyzing. The 2-in. wide collection surface area greatly facilitates division of the collected samples, when different analyses or sample extraction procedures are to be run on the same sample.

Heavy film deposits can be examined by techniques such as infrared absorption spectroscopy, X-ray diffraction, etc. Deposit density changes or composition changes can sometimes be read directly off the film, thereby giving the composition or concentration changes as a function of time within each of the various particle size ranges.

Normally, the particulate matter is removed (or dissolved) from the film for analysis. Water is used, for example, to determine the presence of sulfates or nitrates, and acid, to determine elements such as iron or lead. Background levels for analyses such as these were found to be very low; therefore, very small quantities of material can be detected.

Electron microscope grids can also be attached directly to the drum for subsequent viewing, or for the analysis of individual particles by electron microscope or particulate crystals by electron diffraction. Two-sided sticky tape has worked well for fastening grids onto the drums.

All collection-analysis methods have their limitations. These limitations are determined by: (1) the sensitivity of the analysis procedure; (2) the background level of the component to be detected; (3) the concentration of that component in the aerosol; and (4) the amount of aerosol sampled. An example of the amount of particulate collected is given below.

Example: If ambient air containing $150 \mu g \, m^{-3}$ of particulate matter is sampled for a 24-h period at a flow rate of $5 \, ft^3 \, min^{-1}$, then about 30 mg of particulate would be collected out on the four impactor stages plus the after filter, giving an average of 6 mg per collection surface. If only 1 mg of dirt were required for an analysis, the collection films could be cut up into 4-h intervals giving an average of 1 mg per section.

Atmospheric air sampling results have shown that the cleanout film may have less than 1 per cent of the total dirt and the dirtiest film greater than 50 per cent of the total dirt collected.

RESULTS

During November 1968, the Lundgren Impactor was used to obtain 10 samples of atmospheric particulate matter. These samples, all obtained on the Riverside campus of the University of California, were analyzed as follows:

(1) Particulate weight distribution was determined.

(2) Water soluble fraction was extracted and analyzed for sulfates and nitrates by standard wet chemical methods (JUTZE and FOSTER, 1967).

(3) Nitric acid fraction was extracted and analyzed by atomic absorption spectrophotometry for lead and iron.

Results of these ten tests are plotted in FIG. 3 as micrograms of the particulate, the nitrates, etc., per cubic meter of air sampled for each of the four impactor stages plus the impactor after filter. In all cases, samples were run from 4 p.m. of one day until 8 a.m. of the next. None of these test runs were divided into time fractions; therefore, the numbers given represent averages over the 16-h sampling period.

Data for the ten runs were used to obtain an average distribution for particulate weight, sulfates, nitrates, lead and iron. FIGURE 4 is a log probability plot of these weight averages vs. particle diameter. The plotted diameters are based on assumed density one of spherical particles and an impactor flow rate of 2.9 ft^{3}min^{-1}.

TABLES 1A and 1B summarize the results for ten separate impactor samples describe the type of day on which these samples were taken and tabulate corres ponding averages for wind data, temperature, humidity, and concentration of variou

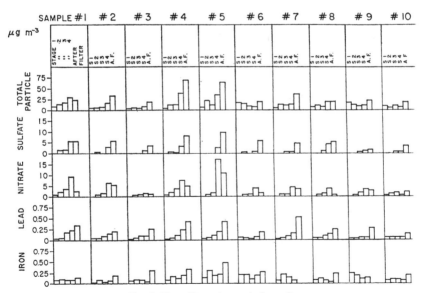

FIG. 3. Plot of impactor analysis results.

23

TABLE 1A. TABULATION OF IMPACTOR ANALYSIS RESULTS PLUS VARIOUS GAS AND METEOROLOGICAL MEASUREMENTS.

Run	Total particulate			Sulfate			Nitrate			Lead			Iron		
	μg m^{-3}	MMD(μm)	σ_g	μg m^{-3}	MMD(μm)	σ_g	μg m^{-3}	MMD(μm)	σ_g	μg m^{-3}	MMD(μm)	σ_g	μg m^{-3}	MMD(μm)	σ_g
1	92	1.0	7	13.1	0.6	4	18.1	1.1	3	0.80	0.6	5	0.53	2.2	10
2	64	0.5	8	9.1	0.3	6	12.8	0.6	3	0.52	0.9	6	0.45	1.1	12
3	40	0.7	16	4.5	~0.15	—	3.7	0.9	3	0.48	0.5	7	0.61	0.6	20
4	144	0.5	7	12.6	~0.15	7	18.0	1.0	4	0.85	0.5	5	0.89	1.1	10
5	141	0.6	10	12.6	~0.15	—	29.9	0.7	2	0.79	0.4	6	1.28	1.7	7
6	69	4.5	10	7.3	~0.1	8	6.2	0.9	3	0.35	0.6	20	0.99	2.2	11
7	79	1.0	8	6.0	~0.15	5	8.6	0.7	3	0.80	~0.2	10	0.49	5.6	3
8	63	1.2	9	10.5	0.5	3	6.4	1.2	3	0.52	0.6	6	0.52	1.7	8
9	75	2.1	15	2.9	0.6	4	7.7	0.7	4	0.47	0.3	15	0.66	9.3	5
10	47	1.7	15	4.6	~0.2	6	5.3	1.8	4	0.30	0.9	20	0.46	1.4	10
Ave. Values	82	0.9	11	8.3	~0.3	4	11.7	0.8	3	0.59	0.5	7	0.69	2.2	8

TABLE 1B. TABULATION OF IMPACTOR ANALYSIS RESULTS PLUS VARIOUS GAS AND METEOROLOGICAL MEASUREMENTS.
AVERAGES OVER 16-h SAMPLING PERIOD

Run	Date/type Day	Temp. (°F)	Relative humidity (%)	Wind velocity (mph)	Wind direction	Peroxyacetyl nitrate (ppb)	Oxidant (pphm)	Hydrocarbon (ppm)	Nitric oxide (pphm)	Nitrogen dioxide (pphm)	Carbon monoxide (ppm)
1	10/31–11/1 Average day, light smog	55	82	2.0	SSW	1.0	0.2	3.0	1.5	2.5	3.5
2	11/3–11/4 Rain—low smog	57	82	2.0	WSW	1.0	0.4	3.0	~1.0	2.0	2.0
3	11/4–11/5 Very clear, no smog	52	81	2.0	SSW	0.4	0.6	3.0	~1.0	1.5	2.3
4	11/5–11/6 Smog—low visibility	50	79	2.0	SSE	3.5	0.75	2.5	1.5	2.7	3.0
5	11/6–11/7 Smog—low visibility	56	47	1.5	SSE	6.0	1.2	2.0	0.5	1.6	3.2
6	11/7–11/8 Windy, clear	63	32	5.5	SW	2.2	1.5	2.0	0.5	1.5	0.5
7	11/11–11/12 Low smog—clear	58	49	1.0	SSE	0.5	1.7	2.5	2.0	2.0	2.0
8	11/12–11/13 No smog—clear	56	~50	2.0	SSW	0.0	1.2	2.0	1.5	1.2	1.6
9	11/13–11/14 Windy—some smog	52	48	6.0	E	1.3	2.1	1.5	0.5	1.1	2.0
10	11/14–11/15 Rain—no smog	52	71	2.5	SSW	0.2	2.3	2.0	~1.0	1.0	0.5

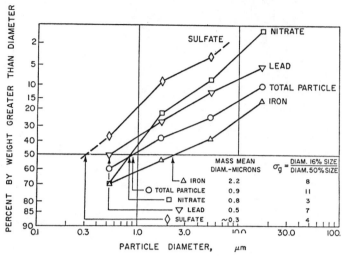

FIG. 4. Average size distributions for 10 impactor samples.

gases present in the air during the sampling period. Estimates of each sample mass mean diameter (MMD) and geometric standard deviation (σ_g), based on an assumed log-normal distribution, are also listed.

All data shown in FIG. 3 are reasonably accurate except for the impactor after filter iron data. This is the only case where the analysis background was consistently high compared to the sample. Although the average iron curve of FIG. 4 should be reasonably good, the individual sample results may have a fairly high error.

CONCLUSIONS

The described sampling instrument is quite simple, but the analysis procedure can be quite demanding. Care must be taken to prevent contamination of the collection surfaces, films must be carefully installed on the collection drums, the impactor nozzles and other internal surfaces must be kept clean and, if used, viscous coatings must be carefully applied or they will tend to run or blow off from the drum collection surface. Once a good technique is developed, determinations of particulate composition, concentration and size distribution can be made from a single aerosol sample.

REFERENCES

LUNDGREN D. A. (1967) An aerosol sampler for determination of particle concentration as a function of size and time. *J. Air Pollut. Control Ass.* **17**, 225–228.
JUTZE G. A. and FOSTER K. E. (1967) Recommended standard method for atmospheric sampling of fine particulate matter by filter media—high-volume sampler. *J. Air Pollut. Control. Ass.* **17**, 17–22.

Shape Factors for Airborne Particles

OWEN R. MOSS

Shape factors for dry aerosol particles generated from 1% solutions by weight of sodium chloride, uranine, or combinations of these two compounds were measured. A factor was obtained for each aerosol, relating particle projected area diameter to aerodynamic diameter, mass, and resistance to viscous flow. The density of dry uranine aerosol particles was calculated to be 1.51 ± 0.12 gm/cm^3, or essentially the same as the material's bulk density.

Introduction

SINCE THE EARLY 1940's, the study and application of aerosols has increased. An important and often time-consuming process is the description of the aerosol used. Without this information, predictions on the system cannot be made.

Geometric and aerodynamic methods are available for describing an aerosol. It is often desirable to predict particle properties in terms of one of these size parameters, once a description of the particle is obtained by the other. Experimental efforts have been made to relate the aerodynamic diameter to the geometric diameter.[1-4] The objects sampled were of irregular shapes and different chemical compositions. As expected, wide variations were observed in the geometric and aerodynamic diameters. In contrast, an aerosol composed of spheres, with the same chemical composition, would be expected to

The work described in this paper was performed under contract with the U. S. Atomic Energy Commission at the University of Rochester Atomic Energy Project, Rochester, New York, 14620.

27

demonstrate a unique relation between the geometric and aerodynamic diameters.

The aerodynamic diameter is an imaginary linear value assigned to each particle to best describe its settling velocity. The value is usually one of two diameters. The aerodynamic diameter, D_a, is the diameter of a unit density (ρ_a) sphere which has the same terminal settling velocity as the particle being observed. All particles, regardless of their shape or density, with the same linear value for D_a will have the same dynamic behavior when the force acting on them is related to their mass.

If the density of the particle is known, a case that is not too common, D_s, or Stokes' diameter, is often used. D_s is the diameter of a sphere with the same density and terminal settling velocity as the particle. It should be noted that both D_s and the density of the particle, ρ, are required to define the particle's aerodynamic properties. Stokes' diameter is not a unique linear indicator of the particle's dynamic behavior.

The commonest geometric parameter is the projected area diameter, D_p. This is the diameter of a circle which has the same area as the image of the particle. The relation of the projected area diameter to other geometric properties has been discussed.[5-9] If the aerosol is not composed of similarly shaped particles, the relation of D_p to any other geometric property will not necessarily remain constant.

Aerosols of sodium chloride and uranine have been used extensively in calibrating aerosol sampling and generating equipment[10-12] An important descriptive parameter for aerosol generation from a solution is the mass distribution of the initially formed wet droplets. This must be predicted from measurements made on the dried aerosol. When the wet spherical droplet is formed, the mass of the resulting dried particle is proportional to the concentration of the mother solution, C'. Knowledge of the mass of the dried particle and the concentration of the mother

TABLE I

Composition of the Mother Solutions

Solution Number	Composition (wt. %)		
	NaCl	Uranine	H_2O
0	1.0	0.0	99
1	0.9	0.1	99
2	0.5	0.5	99
3	0.1	0.9	99
4	0.0	1.0	99

solution would define the diameter of the initial droplet.

The mass distribution of the dry particles is usually defined in terms of geometric or aerodynamic diameters. The density of the individual particle is frequently not known. If the aerosol is composed of similarly shaped members, then the volume of an individual dry particle can be estimated by using a volume shape factor, α_v. For projected area diameter measurements, the volume of a particle of diameter D_p would equal $\alpha_v D_p^3$. If the value $\alpha_v \rho$ is known, the particle mass and the aerosol size parameters of the initially formed droplet can be calculated.

Two factors are then of interest—the factor relating the geometric and aerodynamic diameters of a particle, D_p/D_a, and the factor relating the mass of the particle to its geometric diameter, $\alpha_v \rho$, where $m = \alpha_v \rho D_p^3$. These factors were estimated for five aerosols of different relative sodium chloride and uranine concentrations. The compositions of the various mother solutions are shown on Table I.

Theory

The volume shape factor and density of an individual particle do not readily lend themselves to measurement. For each aerosol studied, it was assumed that these values were approximately constant over the generated size range. The task then became one of calculating an average value from gross measurements. A distribution of particles was collected on a plate. The mass of material collected, M_t, equals the product of the total

29

number of particles present, N_t, and the mass of the particle of average mass.

$$M_t = N_t (a_v \rho D_{\bar{v}}^3) \tag{1}$$

where D_v is the projected area diameter of the particle of average volume. If the collected distribution is log normal, the diameter of the particle of average volume can be calculated from the distribution parameters,[13]

$$D_{\bar{v}} = D_g e^{(1.5 (\ln \sigma_g)^2)} \tag{2}$$

where D_g is the geometric mean projected area diameter and σ_g is the geometric standard deviation. The volume shape factor times density, $a_v \rho$, is estimated from equations 1 and 2.

A relationship between aerodynamic and geometric diameters can be calculated for particles collected according to their dynamic behavior. The motion of an airborne particle under the influence of an external gravitational force and a viscous resistance force is described by a vector relation, where $\vec{V}$ is the

$$m \frac{d\vec{v}}{dt} = \vec{F}_e - \vec{F}_r \tag{3}$$

velocity of the particle, $\bar{F}_r$ is the drag force, and $\bar{F}_e$ is the gravitational force minus the buoyant force.

A spherical particle of unit density settling under these forces will reach a terminal velocity, V_{term}, which can be calculated by approximations of Stokes' law:

$$\vec{F}_e = \frac{\pi}{6} D_a^3 \rho_a \vec{g} \tag{4}$$

$$\vec{F}_r = 3 \pi \eta \vec{v} \frac{D_a}{C_a} \tag{5}$$

$$V_{term} = \frac{C_a D_a^2 \rho_a g}{18 \eta} \tag{6}$$

where $\rho_a = 1$ gm/cm^3, $\eta = $ the viscosity of

the medium, g = the gravitational accelera-
tion, and C_a is the slip correction according
to Davies[14] for a sphere of diameter D_a.

Most particles are not unit density spheres,
and the equation describing the resistance
force must be empirically corrected to allow
for their shape and characteristic geometric
diameter. For projected area diameter measurements on such a particle, the terminal
velocity can be calculated by

$$\vec{F}_e = a_v \, \rho \, D_p^3 \, \vec{g} \tag{7}$$

$$\vec{F}_r = A_r \, \rho \, \vec{V} \, \frac{D_p}{C_p} \tag{8}$$

$$V_{term} = \frac{a_v \, \rho \, C_p \, D_p^2 g}{A_r \, \eta} \tag{9}$$

where A_r = the projected area diameter resistance shape factor, and C_p = the slip correction for a sphere of diameter D_p.

Combining equations 6 and 9 produces the
relation between the aerodynamic and geometric diameter of a given particle:

$$\frac{a_v \rho \, C_p \, D_p^2 \, g}{A_r \, \eta} = \frac{C_a \, D_a^2 \, \rho_a \, g}{18 \, \eta} \tag{10}$$

$$\frac{C_p \, D_p^2}{C_a \, D_a^2} = \frac{A_r \, \rho_a}{18 \, a_v \rho} \tag{11}$$

The ratios of the volume and resistance shape
factors should be constant for an aerosol composed of similarly shaped particles. Equation 11
then implies that the ratio $(C_p D_p^2)/(C_a D_a^2)$
is a constant, z^2. The relation between D_p
and D_a can be derived from z:

$$z = \frac{\sqrt{C_p} \, D_p}{\sqrt{C_a} \, D_a} \tag{12}$$

The function changes as the ratio of the
square root of C_a/C_p.

Experimental Procedure

The aerosol was generated by a Lauterbach generator,[15,16] mixed with dry air, deionized in a tritium deionizer,[17] and collected in a 10-liter chamber. The samples for all the experiments were drawn from the chamber (Figure 1).

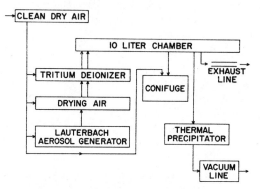

FIGURE 1. The experimental setup.

Volume Shape Factor Times Density, $\alpha v \rho$

The sample was obtained with a thermal precipitator (Thermopositor, Model 100B, American Instrument Co. Inc., 8030 Georgia Ave., Silver Springs, Maryland 20910) which deposited the particles on a 3-inch-diameter No. 2 glass cover slip. A thin-drawn Tygon film (Clear K-83, Carboline Co., 328 Hanely Industrial Center, St. Louis, Missouri 63144) was lowered over the grids to act as a substrate and also to fasten the electron microscope grids to the glass. A layer of chromium was then evaporated onto the plate, grids, and film for conductivity. This assembly of metal, plastic, and glass was then placed on the cold plate of the thermopositor.

The sample was deposited symmetrically with respect to the center of the cover slip, and the plate was divided into 23 annuli, 1/16 inch wide. An electron microscope grid was placed on each annulus such that there was no other grid between it and the center of the plate (see Figure 2). It was assumed

32

that the size distribution on each grid was a good representation of the deposit on the corresponding annulus.

After sampling, the grids were removed and shadowed with chromium. Selected fields about the center of each grid were photographed with an electron microscope, and D_p was measured with a Zeiss particle size analyzer (TGZ-3, Carl Zeiss, Inc., 485 Fifth Avenue, New York, New York 10017). The plate was then washed with a known volume of water. The concentration of the resulting uranine–sodium chloride solution was measured using a Turner Fluorometer (Model 110 Fluorometer, G. K. Turner Assoc., 2524 Pulgas Ave., Palo Alto, California 94303). Incident light of 3650 Å was used, and the intensity resulting from fluorescence was measured for all wavelengths greater than 4000 Å.[18,19] The intensities were compared with those from a calibration solution made from the mother solution; a fresh mother solution was made for each experiment. The mass lost by the removal of the electron microscope grids and the mass that remained on the plate after the first wash were negligible with respect to the total mass on the plate.

Normally, at least 500 to 1000 particles were photographed and counted for each grid. Near the outer edge of the sample plate, the grids representing the deposition on a large area contained very few particles. In these cases, the total number of particles on each electron microscope grid (between 100 and 300) was counted but only four to eight photographs were taken. The assumption was then made that the distribution measured on the photographs was the same as that observed during the counting. In this manner the total number of particles and a rough size distribution could be estimated for these six to eight outer grids.

The size distribution for each grid was corrected to estimate the actual number of particles of a given size increment on the corresponding annulus of the plate. The corrected distributions were summed over the

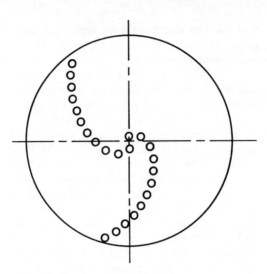

FIGURE 2. Location of the electron microscope grids on the 3-inch-diameter No. 2 glass cover slip.

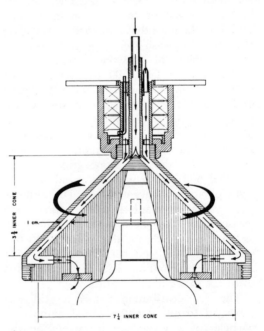

FIGURE 3. Cross-sectional view of conifuge showing flow patterns (from Tillery's paper[20]).

entire area of the sampling plate to give an estimation of both the overall size distribution and the total number of particles.

Relating Aerodynamic and Geometric Diameters

The modified conifuge built by Tillery[20] was used to separate the particles according to their terminal velocities. A sketch of the sampling channel is shown in Figure 3. This machine consists of two coaxial cones. The aerosol is introduced into the space between the cones near the apex of the inner cone. Clean diluting air is introduced at the apex of the outer cone. The cones are mounted on the spindle of a centrifuge and spun clockwise.

The conifuge was designed so that the diluting air could be closely regulated. The total flow through the instrument is determined by the size of the four outlet jets at the base and the rotational speed of the cones. The difference between the diluting airflow and the total required flow through the cones will be the flow through the sampling inlet. To minimize turbulence, it is desirable to have the velocities of the two airstreams as equal as possible at the point where they first come together. To achieve this, the total flow rate through the conifuge at a given rotational speed and the cross-sectional area of the two air inlets at the point of intersection must be known. The flow rates through the machine were measured with calibrated rotameters and a homemade soap bubble volume meter. At 4,000 rpm the total flow rate through the conifuge was 13.8 cm³/sec.

Stöber[21-23] discussed an equation which relates the total flow through the conifuge, the rotational speed of the machine, and the dynamic diameter and density of the particle to some function of the location of the deposit on the outer cone and the physical constants of the sampler. This equation was applied to our conifuge, and the total flow through the machine was calculated. The average theoretical flow was 12.96 cm³/sec, or 6% less than the measured flow rate.

The calibration for the aerodynamic diameter was made by sampling polystyrene latex spheres of density 1.057 gm/cm^3.[24] The distance down the outer cone to the point of deposition was measured for several particle sizes. The aerodynamic diameter for the the spheres was calculated from the density, diameter, and slip factor of the particles. D_a was then plotted against the distance of the deposit down the outer cone. The conifuge was designed so that electron microscope grids could cover the linear distance from the top to the base of the outer cone. From the calibration curve, the center of each grid was assigned an aerodynamic diameter as shown in Table II.

The sampling procedure for each of the experiments consisted of two parts. The first phase was a calibration check. A solution containing several different sizes of polystyrene latex particles was generated into an aerosol in the same manner as described above. This aerosol was then sampled with the conifuge rotating at 4,000 rpm. The second phase consisted of sampling the aerosol of interest.

Results and Discussion

Volume Shape Factor Times Density

The projected area diameter cumulative frequency distribution for each of the experiments did not plot as a simple log-normal distribution. (Figure 4a shows a typical size distribution.) For particles between 0.2 and 1.2 μm, the observed frequency was less than would be predicted from the extension of a straight line drawn through the other 95% of the data (see Figure 4c). This observed distribution may be the actual distribution on the Thermopositor plate, or it may be a deviation from the real distribution due to the measuring procedure for the electron microscope grids near the outer edge of the cover slip.

TABLE II

Comparison of Tillery's and Moss's Data

$$\left(z = \frac{\sqrt{C_p}\, D_p}{\sqrt{C_a}\, D_a} \text{, as measured for a dry uranine aerosol} \right)$$

Grid No.	Conifuge Calibration for 4,000 rpm D_a (microns)	Moss Bp = 74.0 cm Hg	Moss Bp = 73.45 cm Hg	Tillery Bp = 64.10 cm Hg
7	1.7	0.69 ± .11	0.82 ± .15	
8	1.5		0.92 ± .18	
9	1.3	0.64 ± .09		0.69 ± .26
10	1.12		0.82 ± .11	
11	0.98		0.67 ± .07	0.70 ± .43
12	0.84		0.75 ± .08	
13	0.74		0.70 ± .07	0.79 ± .11
14	0.66		0.77 ± .07	
15	0.59	0.55 ± .08	0.70 ± .07	
16	0.53		0.81 ± .07	0.82 ± .08
17	0.48	0.72 ± .07	0.75 ± .12	0.82 ± .10
18	0.43		0.86 ± .09	
19	0.39			0.81 ± .08
20	0.36		0.88 ± .09	
21	0.325			
22	0.3	0.79 ± .10	0.86 ± .07	0.86 ± .07
23	0.275			
24	0.255		0.82 ± .06	0.88 ± .07
25	0.24	0.76 ± .10		
26	0.218		0.85 ± .06	0.89 ± .05
27	0.205			
28	0.195		0.86 ± .06	
29	0.180			
30	0.170		0.84 ± .05	
31	0.160		0.86 ± .06	
32	0.150		0.89 ± .05	0.86 ± .06
33	0.145		0.85 ± .05	0.93 ± .07
34	0.135		0.87 ± .05	
35	0.130		0.88 ± .06	
36	0.125		0.86 ± .06	
37	0.120		0.84 ± .06	
38	0.115		0.86 ± .05	
39	0.110		0.85 ± .05	
40	0.108		0.84 ± .05	
41	0.105		0.91 ± .08	
42	0.101		0.90 ± .08	
43	0.098	0.88 ± .06	0.62 ± .03	
44	0.095		0.85 ± .07	
45	0.092		0.85 ± .07	
46	0.089		0.88 ± .08	
47	0.087		0.80 ± .07	
48	0.086			
49	0.084	0.51 ± .05	0.82 ± .05	
Average			0.80 ± .07	0.83 ± .05

Case 1. Assume that the observed frequency distribution represents the actual D_p distribution. The curve can be accurately described by two log-normal distributions occupying fractions F and $1 - F$ of the sample (Figure 4b). The 'best-fit' values were obtained by successive graphical and computerized iterative processes. The computer ap-

37

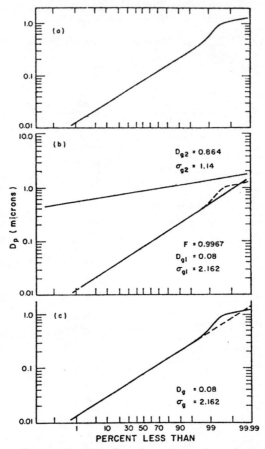

FIGURE 4. Analyzing the volume shape factor data, experiment No. 8. *(a)* The cumulative frequency distribution of projected area diameters. *(b)* Case 1: best fit assuming two log-normal distributions. *(c)* Case 2: best fit assuming one log-normal distribution.

proximations were based on the analytical methods described by Kottler.[25-29] The volume shape factor times density was calculated from

$$a_v \; \rho \; = \; \frac{M_t}{N_t \left(F \; D_{\bar{v}_1}^3 + (1-F) \; D_{\bar{v}_2}^3 \right)} \qquad (13)$$

Should the projected area diameter cumulative frequency distribution be log normal, this method will overestimate the total volume of material on the plate. This would im-

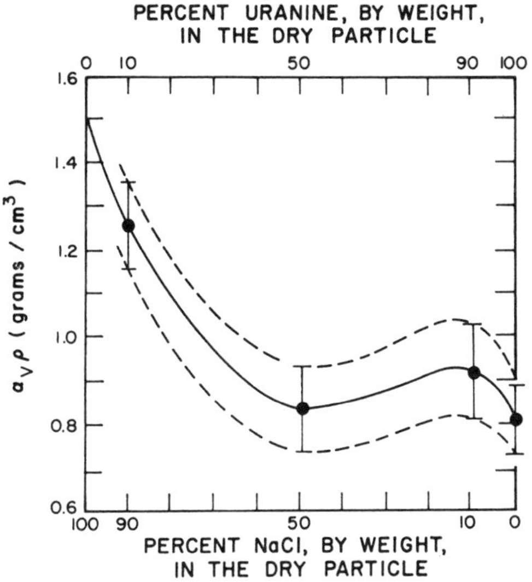

FIGURE 5. Projected area diameter volume shape factor times density.

ply that the calculated values of $\alpha_r\rho$ are too small.

Case 2. Assume a log-normal distribution for the collected data (Figure 4c). If the observed curve for the cumulative frequency distribution is a true description of the sample, then this method will underestimate the total volume on the plate. Consequently, the values of $\alpha_v\rho$ would be too large.

The error in the volume shape factor times density measurement was estimated from the error in each experimental step by progressive summation of simple partial derivatives. One possible source of error not considered in this method was the basic assumption, for the entire measuring procedure, that the deposit on the plate was symmetrical with respect to the central inlet of the thermal precipitator.

The symmetry could be checked only on an empirical basis. After a sample was collected and chromium-shadowed, the plate was placed on a sheet of photographic paper and exposed to light. The faint outline of the

39

TABLE III
Shape Factors for the Five Aerosols

Aerosol Studied:	0	1	2	3	4
Mother solution					
Composition (wt. %) of the dry particle					
NaCl	100	90	50	10	0
Uranine	0	10	50	90	100
Shape Factors:					
$\alpha_v \rho$ (gm/cm³)	1.505	1.25 ± 0.10	0.83 ± 0.10	0.91 ± 0.11	0.80 ± 0.08
$z\left(=\dfrac{\sqrt{C_p}\,D_p}{\sqrt{C_d}\,D_d}\right)$	0.66 ± 0.06	0.76 ± 0.08	0.69 ± 0.08	0.82 ± 0.08	0.815 ± 0.043
A_r $\left(A_r = 18\,\alpha_v\,\dfrac{\rho}{\rho_a}\,z^2\right)$*	11.8 ± 2.1	13.0 ± 2.9	7.1 ± 1.7	11.0 ± 2.6	9.6 ± 1.4
α_v	0.692	0.677	0.609	0.540	0.523
ρ (gm/cm³)	2.165	1.85 ± 0.15	1.36 ± 0.16	1.68 ± 0.20	1.51 ± 0.12

*Solving equation 11.

sample on the negative print was checked for circular symmetry with respect to the center of the plate. If the deposition was noticeably asymmetric, the plate was discarded and another sample was taken. This method could not distinguish depositions on the plate that were slightly skewed toward the half of the plate containing most of the grids (see Figure 2). The deposition of the larger particles on the grids near the outer edge of the cover slip would be most affected by this shift. Half of the outermost six to ten grids would have significantly more particles than the successive grid on the opposite arm of the placement pattern.

Less than 10% of the particles fell on the six outermost grids. According to the laws of thermal precipitation, the larger particles in the distribution will be deposited on these grids. Case 2 considers only a best approximation of the first 95% of the data points. Thus, the area of discrepancy is disregarded, and the results among the pairs of experiments are similar. Case 1 attempts to describe the distribution of the larger particles. It is then more susceptible to this type of sampling error. The best estimate for $\alpha_v \rho$ is the average from the two analytical methods (Table III and Figure 5).

The dry particles of uranine are spheres. Their volume shape factor is then $\pi/6$. If the density of the uranine particle is taken as the bulk density of the material (1.53 gm/cm³),[19,30] the expected $\alpha_v \rho$ would be 0.801 gm/cm³. This is virtually identical with the observed value of 0.80 ± 0.07 gm/cm³. For the case of pure sodium chloride, if all the particles are assumed to be cubes and if each cube lands on a side, the projected area diameter volume shape factor will be 0.692. The bulk density of sodium chloride is 2.165 gm/cm³. This gives an expected $\alpha_v \rho$ of 1.50 gm/cm³.

This project was restricted to aerosols composed of particles with a cubic shape, a spherical shape, or a hybrid shape somewhere between these two limits. The projected area

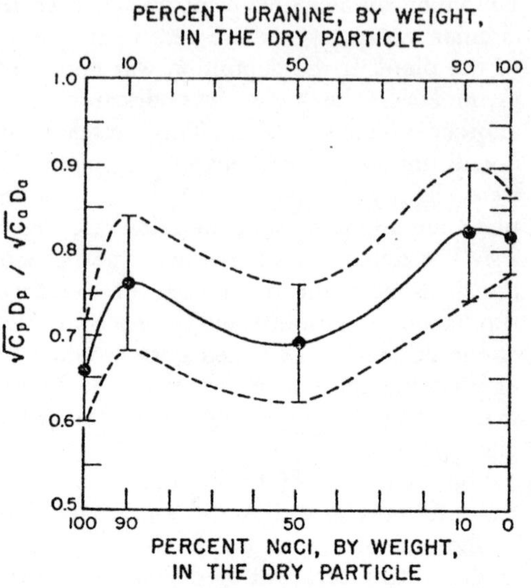

FIGURE 6. Relationship between the geometric and the dynamic diameters.

diameter volume shape factors for a cube and sphere are fairly close. To estimate the density of the particles it was assumed that α_v changes linearly with respect to the percent composition of uranine in the dry particle. The density of the particle can then be estimated (Table III).

Relating Aerodynamic and Geometric Diameters

The main uncertainty in the aerodynamic diameter calibration of the conifuge was the measurement of the distance down the outer cone. The distance of maximum deposition of a calibration aerosol cannot be known to much less than one-half the width of an electron microscope grid, about 1.25 mm. In calculating the experimental deviation, $S_{a(L)}$, to the aerodynamic diameter, $D_{a(L)}$, assigned to a grid located L mm down the side of the outer cone, the following relation was used.

$$S_{a(L)} = \frac{1}{2} \left(D_{a(L+1.25mm)} - D_{a(L-1.25mm)} \right) \quad (14)$$

The experimental error was calculated

from the error in each step by progressive summation of simple partial derivatives. This error in the projected area diameter of the particle of average size was consistently larger than the statistical spread of the size distribution on the grid.

Aerosols generated from all the solutions described in Table I were studied with the conifuge. The results are shown in Tables II and III and Fig. 6. The average of the ratios for each aerosol was taken as a best estimation of z.

$$z^2 = \frac{C_p \, D_p^2}{C_a \, D_a^2} \qquad (15)$$

The dry particles of pure uranine are spherical in shape. The density of a uranine particle should then equal z^{-2} (equation 11). Using the average value of z for the uranine aerosol gives a value of 1.56 ± 0.28 gm/cm³ for the dry particle density.

Tillery,[20,31] using the same instrument, measured the density of uranine to be 1.35 ± 0.05 gm/cm³. Because our calibration curves for polystyrene latex spheres were essentially the same, the difference must lie in the method of analysis. He assumed that the effect of the slip on the particles would not greatly influence the ratio of the dynamic and geometric diameters or the density calculations. This is true within experimental limits for particles of density very close to unity. However, even at densities no greater than that of uranine, the aerodynamic and geometric diameters for submicron spheres begin to differ significantly. The ratio of the slip correction factors for both these diameters then becomes important. In Table II, Tillery's data on uranine, corrected for the D_a and D_p slip factors, are compared with mine. His value for the density of uranine now becomes 1.45 ± 0.18 gm/cm³.

Summary

The projected area diameter volume shape factor times density was defined and measured for four of the aerosols shown on Table III and Figure 5. For aerosol particles of varying compositions of sodium chloride and uranine, $\alpha_v\rho$ varies between 0.7 and 1.5, with the higher value existing for predominantly sodium chloride aerosol particles. The relationship between the projected area diameter and the aerodynamic diameter, D_p/D_a, was shown to be a function of the square root of the ratio of the slip factors, $(C_a/C_p)^{\frac{1}{2}}$. The constant in this relation for each of the sodium chloride-uranine aerosols is shown in Figure 6. Calculations for the projected area diameter resistance shape factor, A_r, were made from the previously discussed equations. The results are shown on Table III.

The density of the dry uranine particle, obtained from the volume shape factor times density experiment, was averaged with the density calculated from the conifuge data including Tillery's work.[20,31] A value of 1.51 ± 0.12 gm/cm^3 is obtained. This is very close to the accepted values reported in the literature.[19,20,30-32] There is no statistical difference between this and the bulk density of the material.

Acknowledgment

I should like to acknowledge the help and patient guidance of Dr. T. T. Mercer in the preparation of this work.

References

1. WATSON, H. H.: Dust Sampling to Simulate the Human Lung. *Brit. J. Ind. Med. 10:* 93-100 (1953).
2. TIMBRELL, V.: The Terminal Velocity and Size of Airborne Dust Particles. *Brit. J. Appl. Phys. Suppl. 3:* S86-S90 (1954).
3. HAMILTON, R. J.: The Relation between Free Speed and Particle Size of Airborne Dusts. *Brit. J. Appl. Phys. Suppl. 3:* S90-S93 (1954).
4. STEIN, F., R. QUINLAN, and M. CORN: The Ratio between Projected Area Diameter and Equivalent Diameter of Particulates in Pittsburgh Air. *Amer. Ind. Hyg. Assoc. J. 27:* 39-46 (1966).
5. DAVIES, C. N.: Shape of Small Particles. *Nature 201:* 905-907 (1964).
6. DAVIES, C. N.: Measurement of Particles. *Nature 195:* 768-769 (1962).
7. DAVIES, C. N.: Size, Area, Volume and Weight of Dust Particles. *Ann. Occup. Hyg. 3:* 219-225 (1961).
8. ROBINS, W. H. M.: The Significance and Application of Shape Factors in Particle Size Analysis. *Brit. J. Appl. Phys. Suppl. 3:* S82-S85 (1954).
9. CARTWRIGHT, J.: Particle Shape Factors. *Ann. Occup. Hyg. 5:* 163-171 (1962).

10. MERCER, T. T., M. I. TILLERY, and H. Y. CHOW: Operating Characteristics of Some Compressed Air Nebulizers. *Amer. Ind. Hyg. Assoc. J. 29:* 66-78 (1968).
11. MERCER, T. T., R. F. GODDARD, and R. L. FLORE: Output Characteristics of Several Commercial Nebulizers. *Ann. Allergy 23:* 314-326 (July 1965).
12. MERCER, T. T., M. I. TILLERY, and M. A. FLORES: *Operating Characteristics of the Lauterbach and Dautrebande Aerosol Generators.* Lovelace Foundation Report —6, Instruments TID-4500 (19th Ed.) (February 1963)
13. HATCH, T., and S. P. CHOATE: Statistical Description of the Size Properties of Non-Uniform Particulate Substances. *J. Franklin Inst. 207:* 369-389 (1929).
14. DAVIES, C. N.: Definitive Equations for the Fluid Resistance of Spheres. *Proc. Phys. Soc. 57:* 259-270 (1945).
15. LAUTERBACH, K. E., A. D. HAYES, and M. A. COELHO: An Improved Aerosol Generator. *A.M.A. Arch. Ind. Health 13:* 156-160 (1956).
16. RAABE, O. G.: The Absorption of Radon Daughters to Some Polydisperse Submicron Polystyrene Aerosols. *Health Phys. 14:* 397-416 (1967).
17. SOONG, AN-LIANG: *The Charge on Latex Particles Aerosolized from Suspensions and Their Neutralization in a Tritium De-ionzer.* University of Rochester Atomic Energy Project. Report UR-49-1000 (1968).
18. GOODMAN, L. S., and A. GILMAN: *The Pharmacological Basis of Therapeutics,* p. 1045, Macmillan, New York (1965).
19. STEIN, F., N. ESMEN, and M. CORN: The Density of Uranine Aerosol Particles. *Amer. Ind. Hyg. Assoc. J. 27:* 428-430 (1966).
20. TILLERY, M. I.: Design and Calibration of a Modified Conifuge. *Assessment of Airborne Radioactivity,* pp. 405-415. International Atomic Energy Agency, Vienna (1967).
21. STÖBER, W.: Design and Calibration of a Size-Separating Aerosol Centrifuge Facilitating Particle Size Spectrometry in the Submicron Range. *Assessment of Airborne Radioactivity,* pp. 393-404. International Atomic Energy Agency, Vienna (1967).
22. STÖBER, W., A. BERNER, and R. BLASCHKE: The Aerodynamic Diameter of Aggregates of Uniform Spheres. *J. Colloid Interface Sci. 29:* 710-719 (1969).
23. STÖBER, W., and H. FLACHSBART: Aerosol Size Spectrometry with a Ring Slit Conifuge. *Environ. Sci. Technol. 3:* 641-651 (1969).
24. PUGH, T. L., and W. HELLER: Density of Polystyrene and Polyvinyltoluene Latex Particles. *J. Colloid Sci. 12:* 173-180 (1957).
25. KOTTLER, F.: Distribution of Particle Sizes. Part I. *J. Franklin Inst. 250:* 339-356 (1950).
26. KOTTLER, F.: Distribution of Particle Sizes. Part II. *J. Franklin Inst. 250:* 419-441 (1950).
27. KOTTLER, F.: The Goodness of Fit and the Distribution of Particle Sizes. Part I. *J. Franklin Inst. 251:* 499-514 (1951).
28. KOTTLER, F.: The Goodness of Fit and the Distribution of Particle Sizes. Part II. *J. Franklin Inst. 251:* 617-641 (1951).
29. KOTTLER, F.: The Logarithmicro-normal Distribution of Particle Sizes: Homogeneity and Heterogeneity. *J. Phys. Chem. 56:* 442-448 (1952).
30. SEHMEL, G. A.: The Density of Uranine Particles Produced by a Spinning Disc Aerosol Generator. *Amer. Ind. Hyg. Assoc. J. 28:* 491-492 (1967).
31. MCKNIGHT, M. E., and M. I. TILLERY: On the Density of Uranine. *Amer. Ind. Hyg. Assoc. J. 28:* 498-499 (1967).
32. LIDWELL, O. M., and D. PHILL: Impaction Sampler for Size Grading Airborne Bacteria-Carrying Particles. *J. Sci. Instr. 36:* 3-8 (1959).

M. Kertész-Saringer
E. Mészáros
T. Várkonyi

TECHNICAL NOTE

ON THE SIZE DISTRIBUTION OF BENZO(a)PYRENE CONTAINING PARTICLES IN URBAN AIR

Abstract—The size distribution of particles containing benzo-(a)-pyrene was measured in a polluted part of Budapest. Particulate matter was collected and fractioned by a cascade impactor backed up by a glass fiber filter to capture unimpacted particles. The benzo(a)pyrene content of samples was determined with thin layer chromatography followed by absorption spectrofotometry. The results of three series of measurements are briefly presented and discussed.

1. INTRODUCTION

BENZO (a) PYRENE is a very dangerous atmospheric pollutant because of its carcinogenic properties. For this reason numerous measurements of the concentration level of this material in urban aerosols have been made, including some in Budapest (KERTÉSZ-SÁRINGER et al., 1969). Very little is known, however, about the size distribution of the atmospheric particles containing benzo(a)pyrene. The present authors are only aware of the work of DeMaio and Corn (1966) who collected the particles by a two-stage elutriator and glass fiber filters. Their results giving the fraction of the benzo(a)-pyrene in two size ranges indicate that more than 75 per cent by weight of this material is in the respirable range ($r < 2\cdot3$ μm). However, the retention of the respirable particles and the desorption of benzo(a)pyrene from the particles which penetrate the lung are also size dependent, thus it is of interest to measure in detail the size distribution of particles containing benzo(a)pyrene with $r < 2\cdot3$ μm.

In this paper the preliminary results of a program for measuring the detailed size distribution of particles containing benzo(a)pyrene in urban air is presented.

2. EXPERIMENTAL PROCEDURE

The sampling method was similar to that used for water soluble substances previously reported by MÉSZÁROS (1968). Atmospheric particles were fractionated in the four stages of a Casella cascade impactor (on clean glass slides) backed up by glass fiber filters (Nagel & Co., Düren MN 85). The collection efficiency of different jets of the impactor was redetermined as described by MAY (1945), because the suction rate employed was higher ($1\cdot7$ m^3 h^{-1} on an average) than that indicated by the manufacturers and because a particles density of $1\cdot5$ g cm^{-3} was assumed. In his original paper May reports a calibration curve measured experimentally giving the collection efficiency of jet impactors as a function of the so-called inertia parameter. The penetration of the filters used was checked by two filters connected in series behind the impactor. Practically no benzo(a)pyrene penetration was found. In normal conditions the particle classification by the sampling device was as follows (the size limits are defined as radii having 50 per cent collection efficiency on the different stages of the impactor):

1 stage	$r > 3\cdot8$ μm
2 stage	$3\cdot8 > r > 1\cdot2$ μm
3 stage	$1\cdot2 > r > 0\cdot46$ μm
4 stage	$0\cdot46 > r > 0\cdot14$ μm
5 stage (filter)	$r < 0\cdot14$ μm

After completion of sampling the tar content of samples on all stages was extracted in the usual way with benzene. The benzo(a)pyrene content of the extracts isolated with thin layer chromatography was determined with an absorption spectrofotometer (Spectro-MOM 201) according to KERTÉSZ-SÁRINGER (1968).

3. RESULTS

Samples were taken on the roof of the National Institute of Public Health of Budapest, which is located in a rather polluted part of the city. Three samplings were made during normal working hours (from 8 to 16 h) during the period September 11 to December 16, 1969. The sampling time of these three series of measurements is given in TABLE 1 with the indication of the sampled air volume.

The weather conditions during these three periods were typical of summer, autumnal and winter situations, respectively.

The solid lines in FIG. 1. show the size distribution of particles containing benzo(a)pyrene obtained in each of the three periods, while the average curve calculated according to LUDWIG and ROBINSON

TABLE 1

Series	Duration	Air volume (m³)
I.	Sept. 11 to Sept. 24	176
II.	Oct. 15 to Nov. 14	227
III.	Nov. 21 to Dec. 16	187

Duration and volume sampled in each of the three series of measurement.

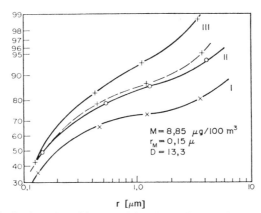

r [μm]

FIG. 1. Size distribution of particles containing benzo(a)pyrene for the three sampling periods (solid lines). The dashed line shows the average distribution. The ordinate gives in per cent the mass of the benzo(a)pyrene associated with particles of radius smaller than r.

(1968) is represented by the dashed line. In the figure the average benzo(a)pyrene concentration (M), the mass median radius (r_M) and the upper decile geometric deviation (D) defined as the ratio $r_{90\%}/r_{50\%}$ are also indicated. It can be seen that the benzo(a)pyrene is distributed on atmospheric particles in a rather large size range. There is some indication that the size distributions are bimodal. The most pronouced difference among these size distributions is the fact that in winter time, when the concentration of benzo(a)pyrene is larger, this pollutant is associated with smaller particles (at least in the size range of $r > 0.15$ μm), and the size distribution has a smaller dispersion than in the summer.

4. DISCUSSION

The results presented here are in good agreement with those of DEMAIO and CORN (1966) showing that the great majority of benzo(a)pyrene, especially in the winter, is in the respirable size range ($r < 2.3$ μm). Our results also show that about half of the atmospheric benzo(a)pyrene is in the range of $r < 0.15$ μm.

An electron microscope study by EINBRODT and AMT (1969) of the particles retained with lungs of people who had died from respiratory diseases showed a similar preponderance of small particles.

47

REFERENCES

DeMaio L. and Corn M. (1966) Polynuclear aromatic hydrocarbons associated with particulates in Pittsburgh Air. *J. Air Pollut. Control Ass.* **16**, 67–71.

Einbrodt H. J. and Amt U. (1969) Die Korngrösseverteilung in Lungenstäuben von Bewohnern aus Geb. mit und ohne Luftverschmutzung. *Arch. Hyg. Bakt.* **153**, 98–104.

Kertész-Sáringer M. (1968) Use of thin-layer chromatography in the determination of 3,4-benzpyrene in dust samples obtained from the air. *Egészségtudomány* **12**, 129–134.

Kertész-Sáringer M., Mórik J. and Morlin Z. (1969) On the atmospheric benzo(a)pyrene pollution in Budapest. *Atmospheric Environment* **3**, 417–422.

Ludwig F. L. and Robinson E. (1968) Variations in the size distributions of sulfur-containing compounds in urban aerosols. *Atmospheric Environment* **2**, 12–23.

May K. R. (1945) The cascade impactor: an instrument for sampling coarse aerosols. *J. scient. Instrum.* **22**, 187–195.

Mészáros E. (1968) On the size distribution of water soluble particles in the atmosphere. *Tellus* **20**, 443–448.

EVALUATION OF AEROSOL POLLUTION USING
A THREE DEVICE SYSTEM

L. MAMMARELLA

ABSTRACT

The characterization of air pollution)in a particular area is connected with the acquisition of a series of statistically significant data. The actual case referred to is a sampling campaign developed in a station in an urban area (Rome), throughout a period of 12 months. For controlling the behaviour of aerosol pollution according to a most complete characterization the three following apparatuses have been employed:

(a) continuous impact sampler (Hirst Spore Trap)
(b) continuous filtration sampler (Aisi Sampler)
(c) differential 4-stage sampler (Mammarella).

Through comparison of data resulting from the three series of sampling it has been possible to analyze:

(i)   the behaviour of pollution in its complexity
(ii) the rate of pollution due to coarse aerosol particles
(iii) the median diameters of different fractions of aerosols.

INTRODUCTION

There are several methods for sampling and classifying air pollutants; in addition the types of samplers, when used to determine both aerosol or gaseous (vapour) pollution, may differ. With regard to aerosol sampling procedures, in particular, the best known and most frequently applied principles are the following: settling, impaction (both on liquids and solids), filtration and the application of thermal or electrical forces.

Settling procedures are principally employed for characterizing coarse aerosols; in fact the more diameters of particles decrease, for a given mass value, the more the settling speed decreases. The collection efficiency for small particles is therefore more and more influenced by meteorological parameters (namely, temperature and wind). Consequently such a type of procedure may be considered useful only for aerosol particle sizes of over 30–40 microns. Higher efficiencies may be obtained however by exposing deposit gauges for long periods of time, utilizing in this case the washing action of rain.

Impaction procedures, both on liquids and solids, exhibit a wide range of efficiencies, which depend principally on the suction speed and the diameter of suction openings. However, in contrast to the settling technique, an impaction sampling implies a volumetric measure of the inspired flow and, as a consequence, one may correlate the pollution levels with the flow volume assayed.

49

Filtration procedures also give different efficiencies, depending on the flow rate or the porosity or thickness of the filter. Frequently in this order of apparatus, large volume samplers are employed. However these devices, while allowing the collection of a relevant quantity of aerosols, are mainly directed to characterize coarse aerosols. Dependence on the fact that small particles are increasingly discharged behind the filter system, necessitates a longer aspiration time for release, scattering phenomena, etc. Finally, thermal or electrostatic precipitators are highly efficient; however, and particularly in the case of the thermal devices, the flow rate is so low that it is very difficult to find a statistically meaningful interpretation.

Above all it is very difficult to foresee continuous sampling over an extended period. The evaluation of aerosol pollution must be extended to all diametric values with the possibility of making long period samplings. Only after fulfilling such conditions will it be possible to obtain useful results. For this reason efficient, but at the same time simple, devices capable of being employed for prolonged periods of time are needed to analyse representative air volumes. In this way sample periods can be extended to days, weeks and months; it must be possible to evaluate both the deposition and the risk of inhalation.

The simplicity of the apparatus described here involves, from a practical point of view, the possibility of employing unspecialized people and hence a low cost of all the operations and devices. These considerations are of prominent interest when we remember that often, for coordinated research work, a high density sampling becomes necessary.

DEVICES EMPLOYED

From the above considerations we recognize the necessity of having more than a single type of collector in a sampling system, because the whole system must be able to identify and classify various aerosol sizes. Apart from providing a characterization of the settling rate and risk of inhalation, it will also be possible in this way to envisage the behaviour of gases and vapours in the atmosphere, the diffusion model of these pollutants being strictly related to that of fine aerosols. In order to have a good sampling response, a three device system as now described has been chosen: an impaction continuous sampler (Hirst Spore Trap), a filtration continuous sampler (Aisi) and a four stage aerosol differential sampler (Mammarella).

The Hirst apparatus (see *Figure 1*) is essentially a suction pump (flow rate 20 litres/minute) and a slit sampler; the aerosol impacts over a microscope slide which moves slowly from the bottom to the top (speed: 2 mm/hour). Along the slide a depositing trace corresponding to the pollution levels through the different periods of a day will become evident. The sampler, which can be employed regularly over periods of months, was developed by Hirst with the primary purpose of sampling airborne spores or pollens (hence the name 'Spore Trap') because it attains a better selectivity for low mass particles of diameters above 5–7 microns.

The Aisi apparatus (see *Figure 2*) may develop a suction speed very near to that of the Hirst sampler. An air stream passes through a paper strip filter automatically moving with a preassigned period. In this way a series of spots is obtained along the paper strip, the optical density of which depends on the quality and quantity of depositing aerosols. In our case we have timed each spot every four hours, thus obtaining six readings each day. Furthermore the Aisi device is able to operate continuously for a very long time.

The four stage differential sampler (Mammarella, see *Figure 3* and described on pages 715–719) is a new type of cascade impactor. Because of the

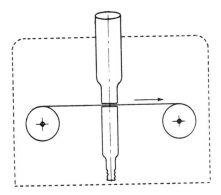

Figure 1. Diagram of the Hirst Spore Trap.

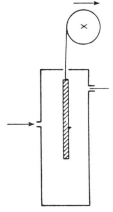

Figure 2. Diagram of the Continuous Aisi Sampler.

length of the sampling strip (practically the length of a microscope slide), the sampler is able to assay medium and high air volumes, subdividing an aerosol into four size classes. This cascade impactor has been employed for discontinuous operations, however, always assaying the same air volume (300 litres at a rate of 30 litres/minute), to control the size variations through the various seasons and through different hours of the same day.

Evaluation of aerosol pollution has been made as follows:

(1) for Hirst and Aisi samplings a densitometer device is used (the same instrument for both apparatuses to obtain the same magnitude of reading).

(2) for the differential aerosol sampler the different slides are observed under a microscope and the deposited particles are classified.

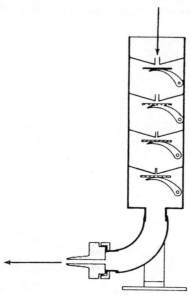

Figure 3. Section through the Four Stage Differential Sampler.

RESULTS

Tabulation of the data has allowed construction of daily curves for the Hirst samplings and six histograms per day for the Aisi samples, which were plotted in a Cartesian axis system. The curves derived via the Hirst apparatus indicate the behaviour of coarse particles, while the Aisi device histograms characterize the entire aerosol pollution.

We carried out experiments for some three years with this method and we have had very interesting results. For example, a winter day sampling in Rome (see *Figure 4* top) shows a noticeable concentration of coarse aerosols (particularly in the early morning hours and in the early afternoon) accompanied by a corresponding quantity of complete (total) pollution. Both coarse and whole aerosols exhibit a certain decrease (not always exactly proportional to each other) on passing from winter, through spring, to the summer months. *Figure 4*, lower portion, shows a sampling result during the spring (March 13) and *Figure 5* shows the minimal levels in summer. By comparing the double series of results (obtained from the two apparatuses) one may derive useful indications regarding the prominent sizes of captured aerosols throughout the various seasons.

A more clean and complete orientation may however be obtained through the simultaneous examination of data derived from the employment of the cascade impactor; for the same time periods of observation, *Figure 6* shows

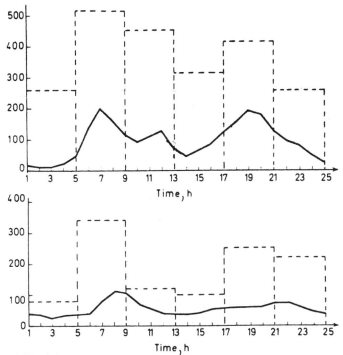

Figure 4. The behaviour of aerosol pollution (histograms: total; coarse aerosols: curves)
Top: a winter day, bottom: a spring day.

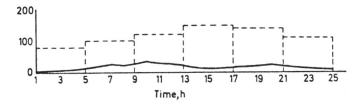

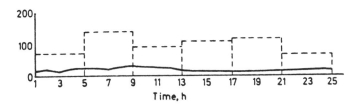

Figure 5. Minimal aerosol pollution levels on summer days (top: June, bottom: July).

the incidence of size classes in a more detailed manner. The continuous decrease of coarse aerosols starting from winter to summer (from about 15–5 per cent for aerosols larger than 20 microns, and from 20–10 per cent for aerosols of about 10 microns) is particularly evident. While the median size particles tend to maintain a constant rate during the different seasons, the smaller particles increase from 20–30 per cent (winter) up to 50–55 per cent (summer).

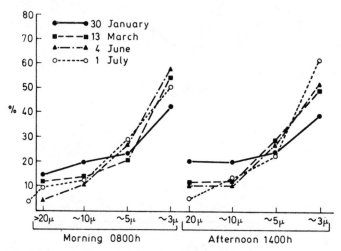

Figure 6. Variation of granulometry of aerosols in different periods of the year (measurements taken with the aerosol differential sampler).

DISCUSSION AND CONCLUSIONS

An examination of the results obtained demonstrates the interesting possibility of using an integrated system of aerosol sampling devices. In fact, a settling sampling system would characterize mainly the coarse aerosol components without the possibility of extrapolating volume concentration data; on the contrary an impaction sampling procedure will give correlated data of the same order but more reliable and complete. If we compare the data obtained from the impaction and filtration systems, it is easy to make correlation between the coarse aerosol pollution and the total pollution. Furthermore the periodic employment of a cascade impactor allows the comparison of diametric values through the different seasons and hours. From an analysis of the slides of the cascade impactor it is also possible to obtain more detailed data for the size, morphology and nature of aerosols.

Finally, another advantage may be the possibility of relating a greater or smaller diffusion of gases or vapours, according to the ratio of fine aerosols. Other pollutants (e.g. biological) may also be studied with the above mentioned system. In fact, by examining the findings of both the Hirst and Mammarella devices under a microscope one may identify numbers and species of spores, pollens, or other biological pollutants. Such a system appears, therefore, to be capable of representing the backbone of a sampling system.

Bibliography

L. Detrie, *Pollutions Atmospheriques* Dunod-Parigi (1969)

P. H. Gregory and J. M. Hirst, *J. Gen. Microbiol.* **17**, 135 (1957).

J. M. Hirst, *Ann. Appl. Biol.* **39**, 257 (1952).

J. M. Hirst, *Trans. Brit. Mycol. Soc.* **36**, 375 (1953).

P. L. Magill, C. Ackley and F. R. Holden, *Air Pollution Handbook*, McGraw-Hill Co. New York (1956).

L. Mammarella, *Inquinamenti dell'Aria, Diffusione e Controllo-Il Pensiero Scientifico Editore* Roma (1968).

L. Mammarella, *Il prelievo differenziale di aerosoli mediante un campionatore a 4 stadi lineari sovrapposti-Annali di Medicina Navale*-LXXI-II, 219–232 (1966).

L. Mammarella, *L'Inquinamento Atmosferico nella Città di Roma. Nuovi Annali d'Igiene e Microbiologia.* **1**, (1969).

L. Mammarella and S. U. D'Arca,: *Variazione Giornaliera ed Oraria dei livelli di pollini e spore fungine nell'aria libera di un grosso centro urbano, Igiene e Sanità Pubblica*-XXII, 11–12; 501–527 (1966).

L. Mammarella and A. Biondi, *L'Accertamento Rapido dell'Inquinamento pulviscolare dell'aria (proposta di un nuovo metodo di indagine). Nuovi Annali di Igiene e Microbiologia* **4**, (1967).

L. Mammarella and L. Di Giambernardino, *Prove di prelievo comparativo fra il Cascade Impactor (May) e il Campionatore Differenziale a 4 Stadi (Mammarella) Nuovi Annali di Igiene e Microbiologia* XVII-5, 435–452 (1966).

Research Appliance Company, *AISI Sampler-Bulletin* 2322. Allison Park, Pennsylvania, U.S.A.

A. P. Stern, *Air Pollution*, Vol. 1 Academic Press, New York (1962).

G. H. Strom, *Atmospheric Diffusion of Stacks Effluents*, 118–194, A. P. Stern, *Air Pollution* Academic Press, New York (1962)

A Multi-Stage Aerosol Sampler for Extended Sampling Intervals

MORTON LIPPMANN, Ph.D. and AGIS KYDONIEUS, Ph.D.*

The design and laboratory evaluation of a new multi-stage aerosol sampler, incorporating an array of six 10-mm diameter cyclones operated in parallel at varying flow rates is described. Each cyclone samples from a common axial inlet tube at a different flowrate, and therefore, has a different characteristic particle size cut-off. The undersized particles are collected on filters. The particle size distribution of the ambient aerosol is determined from the sample masses on the back-up filters and a parallel filter which has no pre-collector. Since the cyclone efficiencies do not vary with loading, the sampling intervals can be as long as necessary to obtain the desired sample masses.

Introduction

A MULTI-CYCLONE SAMPLER for determining the aerodynamic size distribution of airborne particles has been developed. Cascade impactors are presently used for such determinations, but their use is often limited by their inability to collect sufficiently large samples for accurate mass analysis in each size fraction.

Design Objectives

The primary objective was to develop a compact light-weight aerosol sampler which would classify the sampled particles on the basis of their aerodynamic size and collect sufficient sample to permit accurate mass

This research was supported under Grant EC-00231 Environmental Control Administration, Public Health Service, Department of Health, Education and Welfare, and by the American Medical Association Education and Research Foundation. This work is part of a center program supported by the National Institute of Environmental Health Sciences, ES 00260.

analysis and/or chemical analysis for each fraction collected. Furthermore, the sampler should permit easy and rapid removal of collected samples in the laboratory or in the field, and withstand rugged field use and abuse without damage.

Instrumental Approach

Characteristics of Prototype Multicyclone Sampler

The prototype instrument illustrated in Figures 1 and 2 consists of six two-stage samplers operating in parallel and sampling from a common inlet duct. Each two-stage sampler consists of a 10-mm nylon cyclone followed by an efficient filter. The use of this type of commercially available cyclone as the first stage of a two-stage sampler was first described by Lippmann and Harris,[1] who demonstrated that the cyclone retention could be made to closely approximate the particle collection characteristics of the human upper respiratory tract by operating it at an appropriate flow rate.

Unfortunately, there has been a continuing uncertainty as to what the proper flowrate is for respiratory tract simulation. The AIHA Aerosol Technology Committee has recently reviewed available data on this subject and recommended that the proper flowrate for this application is 1.7 liters/minute.[2] Calibration data presented in the paper confirms the validity of this recommendation.

The collection characteristics of the cyclone are flowrate dependent and therefore the cyclone can operate with different particle size cutoff characteristics by operating at different sampling rates. In the multicyclone sampler described herein, this characteristic has been utilized to obtain a series of size cuts by operating the cyclones in parallel at different flowrates. Since each sampler is drawing aerosol from the same sampling tube, the concentration presented to each sampler inlet is the same. However, the ratio between the cyclone collection and backup filter collection will be dependent on the sampling rate.

57

This type of sampler can provide the same type of information as a cascade impactor, but does not have its operational limitations with respect to the mass of sample which can be collected. In an impactor, when solid particles strike previously collected particles instead of an adhesive coated surface, they tend to bounce off and go on to the succeeding stage. Cyclone collection characteristics are relatively unaffected by the amount of sample collected, and the sampler can be run as long as is necessary to obtain adequate sample masses for analysis.

Cascade impactor data are used to construct a size mass distribution curve by plotting the cumulative mass, i.e., the mass on a collection stage plus the masses on succeeding stages against a characteristic size for that stage. In effect, this converts the cascade impactor data into the same type of data obtained directly from the parallel cyclone sampler. For the parallel cyclone sampler, the data would be normalized for the varying flowrates and the percentage on the filter can be plotted against the characteristic cutoff for that stage directly on log probability paper.

Description of Prototype Sampler

The sampler inlet is a 5-inch length of ¾-inch aluminum pipe which brings the sampled air into the center of the housing where it distributes into the six cyclone inlets and also into the inlet of the direct filter sampler at the top center which has no cyclone precollector. Samples entering the cyclone are depleted of larger particles on the basis of the cutoff characteristics of the cyclone, with the smaller particles passing up through the axial exhaust tube of the cyclone directly onto a filter which captures them.

In the prototype instrument shown in Figure 1, the three cyclones with the lower flowrates and the direct filter sampler utilize 13 mm diameter filters supported by plastic Swinney type filter holders. The three cyclones with the larger flowrates deposit their finer particles on 25 mm diameter filters supported

FIGURE 1. Multi-cyclone sampler assembly. The common air inlet is at the left side. The six peripheral filter heads at the right side are each coaxial with a 10-mm nylon cyclone within. The central filter head samples directly from the inlet aerosol.

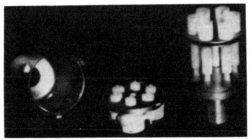

FIGURE 2. Multi-cyclone sampler disassembled. The vortex finder parts of the cyclones which are visible on the inside surface of the cover plate are permanently attached to the filter heads on the opposite side.

by the larger Swinney type filter holders. The cyclones are contained within a half-liter stainless steel beaker which has been modified as shown in Figure 1 by cutting the hole for the sampler inlet in the bottom, and fitting it with three spring latches at the top. The clearance around the center hole is sealed by a gasket between the axial flange of the inlet tube and the base of the can. The space around the sampler inlets which connects with the sampling tube is separated from the ambient air and the rest of the housing by the two O-ring seals which can be seen in Figure 2. The vortex finder parts of the cyclones, which plug into the tops of the cyclone bodies when the assembly is completed, are cemented permanently to the filter heads by epoxy casting resin, which also seals the space between the filter heads and the top plate.

The overall sampler as shown in Figure 1 is about $3\frac{1}{4}$ inches in maximum diameter and 7 inches long. It weighs about $1\frac{1}{3}$ pounds. Therefore, it is small enough and light enough to be used as a hand held sampling head for breathing zone sampling, or possibly even to be worn with a harness on the worker's shoulder as a personal sampler head. The major restriction in its use for personal sampling would be the size and weight of the needed air mover, if it is worn by the worker, or the loss in the worker's mobility if the sampling head is connected to a fixed location suction source by a length of flexible hose.

In the laboratory calibration tests the sampling rates for each two-stage sampler were controlled by a critical orifice between the back side of the filter and a vacuum source. For field use, individual flowmeters could be used, or a series of flow proportioning orifices could be used in conjunction with a total flowrate meter.

Methods

Test Aerosol

The test aerosol used was composed of monodisperse ferric oxide spheres produced by a spinning disc generator which has previously been described.[3] The particles are dried aggregates resulting from the evaporation of the dilute aqueous ferric oxide colloid droplets produced by the generator. The size distribution of samples of the test aerosol were determined by optical microscopic sizing of samples collected on membrane filters which were operated in parallel with the multicyclone sampler. A filar micrometer was used to measure the diameter of 50 randomly selected particles on each membrane filter sample. The geometric standard deviation of diameters in the test aerosols was in all cases equal to or less than 1.10. The ferric oxide particles have a density of 2.5 and conversion of measured diameters to aerodynamic mass median diameter has been made for the data presented in the tables, figures and discussion to follow.

Determination of Collection Efficiency

The actual calibration data to be discussed were collected with a slightly different earlier prototype instrument than that shown in Figure 1. The most significant difference was that the earlier version did not have a whole sample filter at the center of the array, and the reference filter sample was collected on a separate parallel sampling port. The particles were tagged with technetium-99m as technetium sulfide and the cyclone and filter collections were analyzed by gamma counting with scintillation detectors. The samples were measured at the focus of an array of large detectors whose primary application is chest burden measurements of inhaled activity, and which has previously been described.[4] Within a central area in the focus of this array the size or configuration of the sample does not influence counting sensitivities. Thus it was possible to measure the amount collected by the cyclone without removing it from the cyclone, and this greatly simplified the analytical procedure. The measured count rates, after decay and background corrections, were checked for consistency in overall concentration. Data were rejected unless each of the six samplers indicated the same total concentration within plus or minus 10%. In the few instances where such sample rejection was necessary it could usually be traced to plugging of the critical orifice or improper mounting of the cyclone such that leakage was permitted between vortex finder and body of the cyclone.

Results of Comparisons

Collection efficiency was determined as a function of particle size for six different sampling rates. The calibration data are shown in Figure 3 and are summarized in Table I. In these tests, the electrical charge on the particles was neutralized by passing the aerosol over a krypton-85 filled stainless steel capillary which neutralized the average charge per particle from approximately 440 to approximately 40 unit charges. The technique used to measure the average charge is a modification of the technique described by Thomas,[5] and will be described in greater detail in a forthcoming publication. Tests were also run with unneutralized aerosol (Table II) and a comparison of these with the neutralized aerosol tests provides an indication of the effects of electrical charge on deposition in the cyclone. Figure 4 shows the experimental data points for the unneutralized aerosol and the curves from Figure 3 for comparison. It can be seen that the effect of the higher charge level is to increase the cyclone retention, especially at the conditions where the retention is relatively low. On the basis of the comparison in Figure 4, it is likely that the cyclone retention would be somewhat lower than that shown in Figure 3 if the charge level was reduced to less than 40 charges. The curves would probably approach 0% collection at larger particle sizes.

Discussion of Results
Comparison of Results to Previously Published Data

A direct comparison of the calibration data with available data in the literature is possible only at the intermediate flowrates tested. The interest of previous investigators in the performance of the 10-mm nylon cyclone was confined primarily to the determination of the flowrate at which it would reproduce the "Los Alamos" or AEC acceptance curve for first stage collection. As a result, most of the published efficiency data available are at flowrates between 2.8 and 1.2 lpm.

Figure 5 presents characteristic cyclone collection curves for 2.7 and 1.8 lpm from this study, together with data from Ettinger and

TABLE I

Consolidated Experimental Data: Cyclone Retention for Charge Neutralized Aerosols

Sampling Rate liters/min	Weight Percentage Retained in Cyclone at Indicated Unit Density Particle Diameter (Microns)												
	1.21	1.29	2.43	3.32	4.65	5.79	6.65	7.95	8.97	9.28	9.75	11.1	12.5
0.3	5.5	3.3	5.1	7.4	7.7	5.4	11.3	19.4	22.7	*	35.6	59.4	82.3
0.9	5.3	5.7	5.5	5.2	10.8	37.0	46.0	74.1	85.4	91.5	93.0	94.6	97.3
1.8	10.9	9.5	18.0	33.5	72.1	92.8	96.7	*	98.5	*	98.9	99.3	99.5
2.7	29.0	30.0	50.5	78.7	97.5	98.2	98.7	99.3	99.2	99.8	99.7	100.0	100.0
3.5	52.2	55.8	80.7	96.6	98.9	99.4	99.8	100.0	99.9	100.0	99.5	100.0	100.0
4.7	78.8	85.8	97.9	99.5	99.9	99.8	100.0	100.0	100.0	100.0	100.0	100.0	100.0

*Data rejected. Flowrate was unknown due to partial plugging of critical orifice with foreign material.

TABLE II

Cyclone Retention for Electrically Charged Aerosols

Sampling Rate liters/min	Weight Percentage Retained in Cyclone at Indicated Unit Density Particle Diameter (Microns)			
	1.3	3.06	6.05	9.25
0.3	12.6	12.8	30.7	43.4
0.9	13.2	18.9	53.6	89.8
1.8	20.4	60.0	90.5	98.4
2.7	42.9	92.6	98.7	98.7
3.5	59.5	98.4	99.3	100
4.7	80.0	99.7	100	100

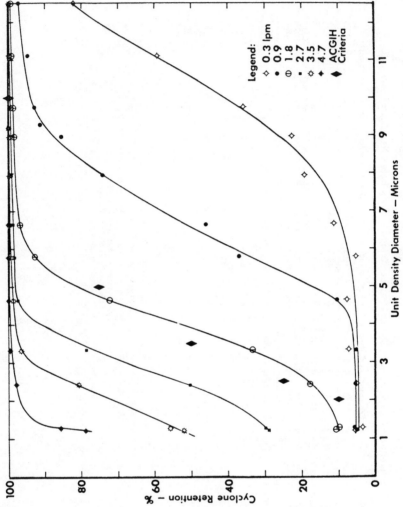

FIGURE 3. Collection efficiency of Dorr-Oliver 10-mm nylon cyclone as a function of particle size at six flowrates.

64

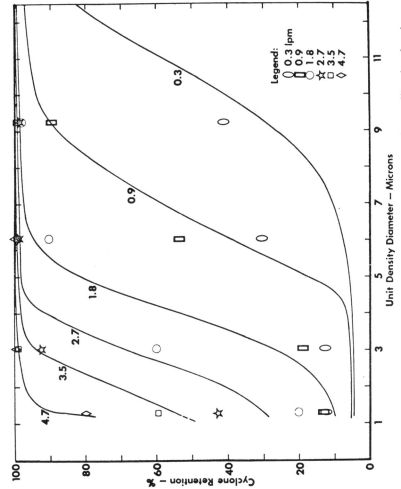

FIGURE 4. Effect of particle charge on cyclone retention. Curves show calibration for charge neutralized aerosol.

65

Royer[6] and Knuth.[7] It is apparent that the data obtained in this work fit extremely well the data obtained by Ettinger and Royer for both 2.7 and 1.8 lpm The data of Knuth show a similar dependence of efficiency on flowrate, but differs with respect to the flowrate required for equivalent performance. In other words, performance closely approximating the AEC and ACGIH first stage sampler criteria takes place at 1.4 lpm according to Knuth,[7] at 1.7 lpm according to Ettinger and Royer,[6] and at 1.8 lpm on the basis of this study. Other data includes that of Sutton,[8] who reported in an oral presentation that the correct flowrate was 2.0 lpm. Tomb and Raymond[9] also recommend 2.0 lpm, but their calibration data was not based on measurement of aerodynamic sizes. Lippmann and Harris[1] recommended 2.8 lpm based on their original calibration. U_3O_8 test aerosols were used and collection efficiencies were determined from projected area particle size measurements of up-and-downstream samples by light microscopy. Kotrappa[10] has shown that by applying appropriate shape factors and converting the Lippmann-Harris measured size data to aerodynamic size the correct flowrate becomes 1.8 lpm.

Use of Multicyclone Sample Data to Determine Particle Size Distribution

Two graphical plotting techniques are commonly used for estimating particle size distribution with cascade impactors. In the MMD method the mass median diameters for the particles collected at each stage are calculated from the number distribution obtained by measurements made with the optical microscope. These mass median diameters are then used as stage constants to obtain the size-mass distribution of unknown aerosols by plotting the sample data on logarithmic probability paper. For each stage, half the mass collected on that stage plus all the mass collected on the succeeding stages (in percentage) is plotted against the mass median diameter for that stage.

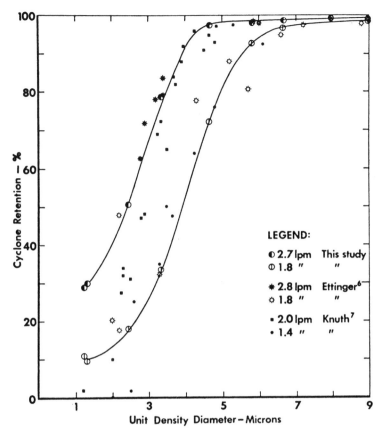

FIGURE 5. Comparison of cyclone retention measurements with data of other investigators.

In the effective cut off method the collection efficiency of each stage is determined as a function of particle size for a test aerosol. The particle size for which the collection efficiency of a particular stage is equal to 50% is then taken as the "effective cut off size" and represents the calibration constant for that stage. The "effective cut off diameters" are then used to obtain the mass distribution of sampled aerosols by plotting the sample data on logarithmic probability paper. The cumulative mass collected on a given sampling stage and on the suceeding stages (in percentage) is plotted against the "effective cut off diameter" for that stage. Mercer has described and compared these two methods very clearly.[11,12,13]

It is apparent that the experimental calibration of the multicyclone sampler is identical to that of the effective cut off method. The effective cut off diameters for the cyclones at six different flow rates can be obtained directly from Figure 3. For the parallel cyclones, the particle size distributions of unknown aerosols can then be estimated by plotting the sample data on log-probability paper. The percentage of sample aerosol which penetrates to each filter stage is plotted against the effective cut off diameter of the corresponding cyclone.

Applications

The multicyclone sampler can be used in various configurations and at various overall flowrates. For instance, the six cyclones could be operated at the six flowrates used in the calibration tests, i.e., between 0.3 and 4.7 liters/minute, with characteristic cut off. diameters from 10 to about 0.5 microns. Alternatively it could be operated at higher or lower flowrates with corresponding shifts in the cut off diameters. Such applications would require additional calibration data beyond the range tested here. At flowrates intermediate to those within the tested range, the cut off sizes could be determined by interpolation of the data.

Another approach would be to reduce the number of different cut off sizes by combining cyclone samples. For example, two cyclones could be operated at the same flowrate and the resulting samples combined for analysis. This would reduce by one the number of independent measurements. If the two lowest flow cyclone samples were combined, it would increase by a factor of two the size of the smallest sample collected.

In field applications only the filters need be analyzed. Since the total concentration can be determined from the filter which operates without a cyclone precollector, it is unnecessary to determine the first stage collection in the other parallel samplers. The fractional removal in each of these two-stage samplers can be determined from the amount on the filter, the flowrate through the filter, and the concentration of the overall sample. Thus, this sampler overcomes one of the major objections to the use of two-stage cyclone samplers, which is the difficulty in recovering sample from the cyclone precollector.

Summary and Conclusions

A multicyclone sampler designed to be conveniently used to obtain particle size distribution of particulate matter and to take sufficient sample for chemical analysis of the fractions has been described. The characteristic curves for the six cyclones were presented and the methods used to determine the particle size distributions from experimental data described. It is shown that the "Effective Cut Off" method used with cascade impactors can be directly used with the multicyclone sampler for estimating the particle size distribution of unknown aerosols.

The sampler is calibrated for particles with unit density diameters between 1 and 12 microns. Aerosol samples with a high fraction of particles below 1 micron could be classified by the sampler if it were operated at higher flowrates than those used in the calibration tests.

It is anticipated that with the availability of this instrument much more size information will be collected in the different industrial and municipal control operations. In addition the availability of this instrument with its capability of collecting large samples, should make it easier to perform chemical analysis of specific size fractions of particulate matter.

Acknowledgment

Grateful acknowledgment is hereby made to Paul Goldwasser, who performed the optical microscopic sizing and aided in the performance of the experimental work.

Appendix

Determination of Aerosol Size Distribution Using Calibration Curves of Individual Samplers in a Multi-Sampler Array

Consider an aerosol containing particulate matter of unknown particle size distribution. When the aerosol is drawn through the multi-cyclone sampler each cyclone will retain a different weight fraction of particles which is dependent only on the characteristic curve of the cyclone and the unknown particle size distribution. Assume that the percent retained on each cyclone has been obtained by standard techniques and is given by Z_1, $Z_2 \ldots Z_6$ respectively. If the abscissa of Figure 3 is divided in six equal segments then six linear equations can be written

$$Z_i = \sum_{j=1}^{6} (x_{ij}) V_j \quad \text{(1) for } i = 1 \text{ to } 6$$

where

x_{ij} = percent retention in cyclone i of particles with diameter between D_{j-1} and D_j

V_j = weight percent of particles in sample population with diameters between D_{j-1} and D_j

Since the z_i's can be experimentally determined and the x_{ij}'s obtained directly from Figure 3, the $V_j's$ of the unknown sample population can be calculated by solving the six equations simultaneously.

It should be noted that equation 1 is only approximate and based on the assumption that particles with diameters in the range D_{j-1} and D_j have percent cyclone retention of x_{ij}. This of course is only true if the characteristic curve is a straight line in the range D_{j-1} and D_j and if the weight fraction of particles of the sample population is the same for all diameters between D_{j-1} and D_j. A more general approach was used by Geisel[14] who obtained a similar set of linear equations.

Both of the above mentioned techniques

were used in an attempt to determine the particle size distributions of unknown sample populations. Both methods failed for the multicyclone sampler because ill conditioned matrices were obtained. The reason was that the characteristic curves were nearly parallel so that the six linear equations were not completely independent of each other.

References

1. LIPPMANN, M., and W. B. HARRIS: Size-Selective Samplers for Estimating "Respirable" Dust Concentrations. *Health Physics 8:* 155 (Mar. 1962).
2. AIHA Aerosol Technology Committee: Guide for Respirable Mass Sampling. *Amer. Ind. Hyg. Assoc. J. 31:* 133 (Mar.-Apr. 1970).
3. LIPPMANN, M., and R. E. ALBERT: A Compact Electric-Motor Driven Spinning Disc Aerosol Generator. *Amer. Indu. Hyg. Assoc. J. 28:* 501 (Nov.-Dec. 1967).
4. LIPPMANN, M., and R. E. ALBERT: The Effect of Particle Size on the Regional Deposition of Inhaled Aerosols in the Human Respiratory Tract. *Amer. Indu. Hyg. Assoc. J. 30:* 257 (May-June 1969).
5. THOMAS, J. W., and D. RIMBERG: A Simple Method for Measuring the Average Charge on a Monodisperse Aerosol. *Staub* (English Translation) *27:* 18 (Aug. 1967).
6. ETTINGER, H. J., and G. W. ROYER: Calibration of a *Two-Stage Air Sampler.* LA-4234 Los Alamos Scientific Laboratory (Aug. 1969).
7. KNUTH, R. H.: Recalibration of Size-Selective Samplers. *Amer. Indu. Hyg. Assoc. J. 30:* 379 (July-Aug. 1969).
8. SUTTON, G. W.: Calibration of Aerodynamic Size vs. Efficiency for a Miniature Cyclone. Presented at the 1966 Annual Meeting of AIHA, Pittsburgh, Pa. (May 1966).
9. TOMB, T. F., and L. D. RAYMOND: Evaluation of the Collecting Characteristics of Horizontal Elutriator and of 10-mm Nylon Cyclone Gravimetric Dust Samplers. Presented at the 1969 Annual Meeting of AIHA, Denver, Colo. (May 1969).
10. KOTRAPPA, P.: Revision of Lippmann-Harris Calibration of Two Stage Sampler Using Shape Factors. Submitted to *Health Physics.*
11. MERCER, T. T.: On the Calibration of Cascade Impactors. *Ann. Ocup Hyg. 6:* 1 (1963).
12. MERCER, T. T.: Aerosol Production and Characterization: Some Considerations for Improving Correlation of Field and Laboratory Derived Data. *Health Physics 10:* 873 (Dec. 1964).
13. MERCER, T. T.: The Interpretation of Cascade Impactor Data. *Amer. Indu. Hyg. Assoc. J.* 26:236 (May-June 1965).
14. GEISEL, W.: Calculating the Particle Size Distribution of a Dust by Means of Fractional Separation Efficiency Curves and Total Efficiency Curves. *Staub* (English Translation) *28:* 25 (Aug. 1968).

A Portable Counter and Size Analyzer for Airborne Dust

W. J. Moroz, V. D. Withstandley, and G. W. Anderson

INTRODUCTION

RECENTLY, considerable attention has been directed toward dust conditions in mine atmospheres which may build up to concentrations posing explosion or health hazards. Both the total count and the size distribution of the particles are significant in determining the hazard potential. To provide the necessary data for rapid analysis of a dust laden atmosphere, a portable continuously operating, immediate readout device, capable of monitoring high particulate concentrations, has long been needed.

The portability, independence of an external power source, and immediate readout capability of the counter and size analyzer, designed and built at the Center for Air Environment Studies, enable it to function with advantage over the presently employed much slower or more cumbersome methods of monitoring mine atmospheres. The instrument can also be used for air quality surveys and for monitoring in industrial mills, foundries, quarries, factories, and urban atmospheres.

The instrument under development operates on the basis of (visible) light scattering by individual particles. The unit is designed to detect, count, and size-discriminate solid particles for at least two ranges between 0.5 and 10 μ. It is a completely self-contained unit whose power source consists solely of rechargeable dry cells. It is portable (total weight including batteries and all components is about 7 kg) and simple to operate. Total power consumption, mainly needed to pump sample air and surrounding filtered "sheath" air at moderate speed past a high intensity light-field volume, is less than 33 W.

Dust particle detecting and analyzing devices based upon the principle of light scattering by individual particles moving through a high intensity light-field volume (also called optical volume, view volume or sensing zone) have been developed by several investigators.[1-4] Later work includes the particle size analyzer developed by Thomas[5,6] at the Southern Research Institute and an instrument built by Sinclair.[7] Several instruments of excellent quality are now available commercially. Among these, the instruments of Bausch & Lomb,[8] Royco,[9] and Southern Research Institute are prominent. These instruments have not been adopted for routine measurements in mines because of the stringent power and weight limitations imposed by practical considerations.

The CAES instrument is being developed to meet the following primary specifications:

(1) true portability; i.e., a size and weight convenient for carrying in a mine. The CAES prototype weighs approximately $6\frac{1}{2}$ kg and a business briefcase contains the entire instrument, including batteries and probe.

(2) independence of external sources of power; i.e., "cordless" operation with complete dependence upon batteries for all power requirements.

(3) fulfillment of mine safety requirements; i.e., design passed "permissible" for use in gassy atmospheres. Ruggedness of construction is essential.

(4) capability of dealing with concentrations up to 1500 particles/cm³ with instantaneous readout.

(5) capability to size discriminate in several size ranges.

I. DESIGN CONSIDERATIONS

A. Electronics

To size and count individual particles continuously by an electro-optical method requires a transducer to convert light energy scattered from the particles into electric signals; discriminator circuits to distinguish pulse amplitudes; gating and counting circuits to record the numbers of pulses in the selected amplitude ranges; indicators and/or recorders to provide readout; and power supplies.

The transducer chosen for this instrument is a 1P21 photomultiplier tube (PMT), which converts a light pulse (scattered from a particle) into a current pulse. The amplitude of the current pulse from the PMT is related

73

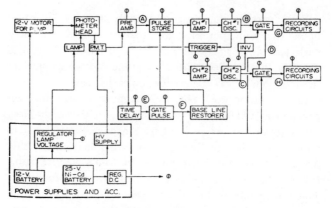

FIG. 1. Block diagram showing electronic components of CAES particle analyzer. The dashed line encloses the power supply.

to the size of the particle while the frequency of pulses is a measure of the concentration of particles. The pulses must be separated into groups according to amplitude and the number of discrete pulses in each group must be continuously integrated. A particle number distribution with respect to particle size is thus obtained for an aerosol.

The block diagram (Fig. 1) shows how these objectives are accomplished for an instrument counting particles in two size ranges. The raw pulse is fed into the preamplifier and the amplified pulse charges a capacitor to an extent proportional to the peak amplitude of the pulse. The voltage decay of the capacitor is negligible over the pulse analysis period. The capacitor thus acts as a pulse signal storage mechanism. A time delay circuit (500 μsec) is incorporated to provide adequate time for a particle signal to reach maximum amplitude. The capacitor voltage is applied, through a voltage divider network, to the channel amplifiers, the relative input to each of which is adjustable via the divider network. When the output voltage of the channel I amplifier is adjusted to be x times larger than that of the channel II amplifier, it follows from the identity of the two "triggers" that the input voltage required to activate channel II is x times that needed to activate channel I. In this way one can adjust the size cutoff point at which channel II is activated and channel I is inactivated. When the input voltage to the channel I discriminator is high enough to activate the circuit, a trigger pulse is applied to the time delay circuit, which is set for

500 μsec. During this interval the channel I discriminator circuit remains activated and the channel II discriminator circuit may or may not be activated. The output of each discriminator is then applied to its respective gate circuit. In order that counting be accomplished in only one circuit, an "override" signal from the channel II discriminator circuit is applied to the channel I gate if the channel II discriminator circuit is activated. Thus the pulse passes through gate I or II depending on the state of the discriminators.

Figure 2 provides a representation for the cases described above. Following channel selection, the signal pulse triggers a multivibrator of constant amplitude and period. The output of the multivibrator is next applied to an amplifier which acts as a buffer as well as the final power

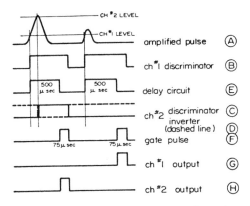

FIG. 2. Representation of channel gate configurations showing signal pulse, time delay, gate pulse, and channel outputs.

amplifier for the recorder or indicator. This constant amplitude, constant width pulse train is integrated by the meter or recorder action, providing an analog readout proportional to the average period of the pulses. At the end of the 75 μsec gate pulse the baseline is restored by discharging the input capacitor in the pulse storage section. The discharge brings all circuits to their previous "waiting" positions. The circuits are then ready to receive and process the next pulse.

B. Optics

As with all instruments operating on the principle of light scattering from individual particles, the critical region of the system is the small volume in which light of high intensity illuminates the particles passing through it. A particle passing through this light field volume gives rise to a pulse of light which will, in part, be scattered in the direction of the photosensitive element, this being the cathode of the 1P21 photomultiplier tube. It is apparent that the intensity of the scattered energy must be sufficient to provide a discrete signal above noise background if the pulse is to be discriminated.

In order that a discrete sharply defined pulse of light scattered from an individual moving particle may be obtained, the field volume itself must be sharply defined; blurring at the edges, halos, vignetting, and other optical defects must be suppressed to a minimum. Further, a level and flat topped pulse can be obtained only if the light intensity is uniform throughout the field volume, and if, at the same time, the scattering particle is spherical or maintains its orientation as it moves through the light field volume. Finally, in order to ensure that the particles pass directly through the field volume and do not pass along or outside of its periphery, it is essential that the axis of the needle centrally intersect the field volume (see A in Figs. 1 and 5).

The feature of viewing the scattering at an angle of 135° (as in the CAES instrument) with respect to the incident beam is by no means as theoretically promising as viewing at 180°. Summaries of the theory of scattering and of the angular distribution of scattered intensity are given by

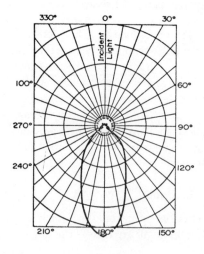

FIG. 3. Angular distribution of intensity of light scattered by a spherical particle of about 1.3 light wavelengths in diameter ($d/\lambda = 1.3$).

Cadle,[10] Davies,[11] and Van de Hulst.[12] Briefly, when a particle is comparable in size or is large relative to the wavelength of incident light, the radiation scattered in a given direction is described by Mie theory. The scattered flux is the resultant of light waves originating from various parts of the particle; among other factors it varies with the index of refraction of the material of the particle and with the ratio of particle size to wavelength. Flux intensity scattered at an angle θ from the direction of the incident beam (taken to be 180°) is described by $I(\theta) = 1/I_0 d[\phi(\theta)]/d\omega$, where I_0 is flux projected orthogonally on a particle, $d[\phi(\theta)]/d\omega$ is flux scattered per steradian at angle θ, and ω represents solid angle. This definition is valid only for illumination which is parallel, unpolarized, and laterally coherent.

Theory indicates that the variation of $I(\theta)$ with θ becomes increasingly lobed in the forward direction, $\theta = 180°$, as the ratio d/λ increases, where d is particle diameter and λ is wavelength of incident light. The effect is predominantly at the expense of the backscattered flux $\Phi(0)$, which, for values of $d/\lambda < 1$, is comparable to $\Phi(180°)$. Figure 3 illustrates the pattern of $I(\theta)$ for light scattered by a spherical particle of about 1.3 wavelengths in diameter and having an index of refraction of 1.5. A marked feature of this pattern is the predominance of scattering in the forward direction.

In the instrument described here, the direct beam from the lamp in the forward direction would present to the PMT a light flux background which would completely mask the light pulses scattered by the particles. On the other hand, since viewing as close to 180° as possible is desirable, for the reason noted above, a compromise between the practical and the ideal led to a viewing angle of 135°.

It is recognized that the response of the detection and sizing system is a function also of other physical properties of the sampled particles, most notably particle composition, shape, and surface properties. The variety and randomness of such factors likely to be encountered lead, from a practical standpoint, to carrying out primary size calibration using spherical particles of known characteristics (polystyrene latex).

C. The Particle Coincidence Problem

The CAES instrument is capable of detecting, pulse

height analyzing, and counting an individual current pulse from the PMT in approximately 575 μsec. To avoid counting two particles as one within this period, the minimum tolerable particle separation in space at the designed flow velocity is $s = t_e V = 0.17$ cm for the prototype, where s is minimum tolerable spacing between particles if pulse coincidence is to be avoided within the processing time, t_e is total processing time (5.75×10^{-4} sec), and V is aerosol velocity through the needle (300 cm/sec). Figures given in parentheses are for the CAES prototype analyzer.

From the standpoint of detection, two particles are "coincident" when they are closer together than the critical distance s. They will then give rise to only a single pulse which will be registered as a single particle. It is because of such possibility of "overcrowding of the particle speedway" that dilution of the sample becomes necessary at high concentrations. To analyze this possibility more fully one must calculate the probability of occurrence of particle coincidence within the light field volume or equivalently inside the sample needle, since all particles are assumed to pass through the field volume at the same speed as that of the air in the needle.

For purposes of analysis, coincidence is defined as the event of two or more particles being closer to each other than s during passage through the field volume. Given that the particles move along at the same speed, the coincidence probability is a function of the particle concentration in the sampled fluid. A coincidence probability in excess of, say, 10% for the duration of any period of sampling is to be avoided. It should be noted here that in the CAES instrument the transit time of each particle through the field volume is about 500 μsec, but the total processing time for an individual particle signal is 575 μsec, the additional 75 μsec being required for pulse height analyzing and counting.

To determine the coincidence probability one can regard the particles (assumed all to be moving at the same constant speed) ideally as points randomly distributed along a straight line. This idealized analog is cited by Parzen,[13] where it is given the title, "The probability of an uncrowded road." By analogy it is assumed that

(1) the particle diameters are small relative to average particle separation so that the particles will not interfere with one another at small center-to-center separations and

(2) the inside diameter of the needle is small compared to its length so that the probability that two or more particles will be abreast of one another within the needle can be neglected because of its restricted lateral dimension.

These two conditions are realized in practice, for the average distance between particles for the dust concentrations of concern is about $10^4 \bar{d}$ and the length of the needle is more than 100 times its inside diameter. The self-evident requirement that the inside diameter of the needle be large relative to particle diameter is clearly realized; the needle diameter is 8.1×10^{-2} cm whereas detectable particle sizes range from 5×10^{-5} to 10^{-3} cm.

The coincidence probability problem and its solution may now be stated in terms applicable to our situation. Along a straight path of length L are N distinguishable particles distributed at random. The probability that no two particles will be less than a distance s apart, for s such that $(N-1)s \leq L$ can be shown to be equal to

$$P_N = [1 - (N-1)s/L]^N. \qquad (1)$$

In our case, N is the number of particles within the sample needle of length L (10 cm) at a given time, and particles separated by a distance less than s are considered coincident ($s = 0.17$ cm).

Substituting the appropriate values into Eq. (1), one obtains a no-coincidence probability of about 90% for $N = 3$. This implies a *mean* separation within the needle of 3.3 cm.

In two dimensions the solution may be demonstrated graphically as shown in Fig. 4. Let the abscissa of a Cartesian coordinate system represent all possible, equally likely positions of the first particle, and let the ordinates represent all possible, equally likely positions of the second particle. Assuming that the particles are independently and randomly distributed along the path length L, the shaded strip represents all possible positions of both particles wherein coincidence can occur. The probability of coincidence of the particle pair is then the fraction $P(c) = A_s/A_t$, where A_s is the area of shaded strip representing all possibilities, and A_t is the area representing all possible simultaneous positions of the two particles.

The area of the strip representing coincidences is $L^2 - 2(L-s)^2/2$ and the area representing all possible simultaneous positions of the two particles is clearly L^2.

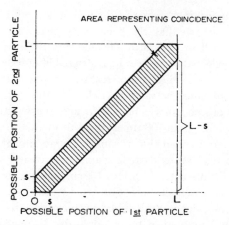

FIG. 4. Positional possibilities for two particles on a line of length L.

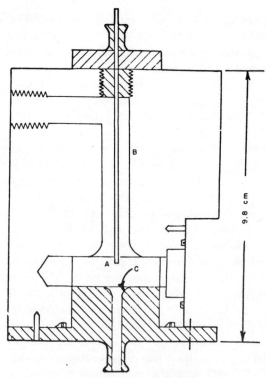

FIG. 5. Photometer head assembly for prototype
CAES particle analyzer.

Thus,

$$P(c) = 2s/L - (s/L)^2. \tag{2}$$

It is seen that Eq. (2) agrees exactly with Eq. (1) after taking into account that P_2 from Eq. (1) refers, of course, to a probability of no coincidence for two particles. That is, $P(c) = 1 - P_2$. Extension of this geometrical version to more than two dimensions (i.e., to many particles) is precisely what the continuously distributed random multi-variate theory described by Parzen accomplishes.

D. The Flow System

The photometer head chamber and air passages are designed to suppress turbulence within the chamber, which operates at pressures slightly below atmospheric. Special features are the bell-mouth sheath-air entrance hole and the bell-mouth total-airflow exit hole in the bottom plate (Fig. 5). Likewise, the position of the sheath-air entrance hole relative to the bend in the sheath-air inlet duct was determined from aerodynamic considerations. The effectiveness of these features in suppressing turbulence or recirculation of the dust laden air in the CAES instrument was confirmed by observing smoke flow patterns and subsequently by noting the absence of particle deposition in the chamber after several hours of operation.

II. DESCRIPTION OF THE INSTRUMENT

Figure 6 is a cross sectional plan view of the prototype CAES photometer showing details of construction with extreme light rays indicated. The light source is a GE 425 filament bulb lighted via the regulated 5 V section of the power supply. The bulb requires about $2\frac{1}{2}$ W and is a source of visible white light.

An image of the filament of the lamp D is formed on the aperture E (Fig. 6), which in turn is collimated and then brought to a sharp focus at A within the light field volume. Light scattered from particles passing through A is focused upon aperture F, whence it proceeds to the photo-cathode G of the photomultiplier tube. Provision is made for adjustment of the position of the phototube so that the light pulses from A will fall upon the most sensitive area of the cathode plate G. Figure 7 is a photograph of the CAES instrument with the arrow pointing to the

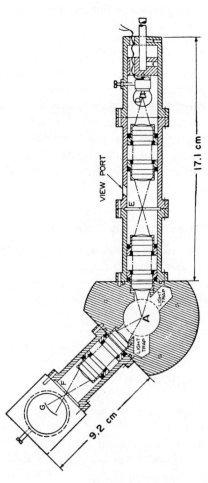

Fig. 6. Cross sectional plan view of the prototype CAES particle analyzer–photometer assembly with extreme light rays indicated.

82

photometer head housing the critical light scattering zone.

All metallic components of the photometer are black-anodized aluminum to reduce glare. The design of the lens system, and of the source and photopickup positioning tubes is similar to that of a miniature laboratory photometer built in 1958 by a group under the direction of V. R. Mumma, but is modified in several ways to satisfy specifications. The photometer head is shown as an assembly drawing in Fig. 5. All lenses are Hastings triplets, anti-reflectivity coated.

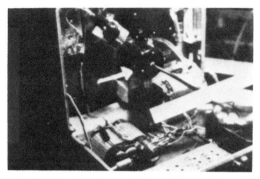

Fig. 7. Photograph of assembled particle analyzer. The arrow points to the photometer head.

The photomultiplier tube (G in Fig. 6) converts a light pulse scattered from a particle in the view volume (A in Fig. 5 or 6) into a current pulse, which is electronically amplified, height analyzed, and continuously integrated with other pulses belonging to the same amplitude range. The integrated final output energizes a recorder or meter which is calibrated in particle number per cubic centimeter for the preselected size ranges.

Power for the motor driven vacuum pump, for the PMT high voltage supply, and for the lamp is supplied by a 12 V battery pack. A separate 25 V Ni–Cd battery supplies the electronic circuits. Regulators are provided in the PMT–voltage supply and in the lamp–voltage supply to provide constant voltage for these components over an input range of 12 to 6 V as the batteries discharge. The regulators permit satisfactory operation of the instrument until the batteries can no longer supply the power necessary to move the required air volume. In Fig. 1 the power supply layout is enclosed by dotted lines.

The sample air containing the aerosol to be analyzed

passes into the photometer head through the hollow needle (Fig. 5). Filtered air (sheath air "envelope"), at the same velocity as that carrying the aerosol, passes through the larger channel B surrounding the needle. The function of the sheath air is to limit turbulence at the edge of the sample stream, to "contain" the sample stream as much as possible until it passes through the shaped exit hole at C, and in particular to minimize the possibility of aerosol recirculation inside the photometer head. In the CAES instrument the flow rate of the sample air is 100 cm^3/min. In order that the sheath air, which passes through a large diameter duct (0.96 cm), have the same velocity as that of the sample air, the sheath-air flow rate must be 12 700 cm^3/min. This flow volume, at a pressure drop of about 0.8 Torr, is close to the capacity of the vacuum pump, which must remain small and light for minimum power and weight requirements.

The miniature 312 g vacuum pump is of the rotary sliding vane type and is designed to move 18 000 cm^3/min running open (with no pressure drop) at 6000 rpm.

A diagram of the flow system is shown in **Fig. 8**. The instrument has a built-in rotameter (3) for measurement of total flow. The total flow through the photometer head is regulated to 12 800 cm^3/min by a bleed valve (5). When this flow rate is achieved, the pressure differences through the sheath-air channel and through the sampling needle are identical, and the same flow velocity is achieved in both passages. An orifice valve (2) is installed in the sheath-air passage to obtain the required resistance for flow through this channel. With the orifice valve correctly set, the flow velocity through the sampling needle is measured under the conditions of operation and the total flow adjusted to provide the correct sheath-air velocity.

III. CALIBRATION

Electronic calibration in terms of number of particles per cubic centimeter for each size range is accomplished in the laboratory by means of an electronic pulse generator. By letting the output of the pulse generator serve as input to the discriminator circuits the readout indicator is calibrated for count. The input voltages to the pulse height discriminators are appropriately attenuated for various pulse amplitudes simulating the output of the PMT.

To date, relative calibration using aerosols has been done through correlation of readout of the CAES instrument with that of commercial instruments (Bausch & Lomb, and Royco). Good agreement was obtained. Absolute calibration by comparison of counts of polystyrene spheres on filter paper (by microscope observation) with CAES instrument results is in progress. Data from these tests will provide information on counting accuracy for the selected size ranges and at the same time will permit identification of the precise upper and lower particle size limits for each range.

Size calibration is achieved using spherical particles of known size and index of refraction. Particles naturally vary in their reflecting properties because of a variety of other characteristics, such as differences in shape, material composition, and surface state, and some differences in instrumental response, indeed occurs as a result. From the practical standpoint, a standard for the "nonideal" particle size is best and most frequently expressed in terms of sphere equivalent diameter. Monodisperse aerosols, consisting of polystyrene spheres of known diameter (size distributed with standard deviation <0.01) generated in the laboratory, constitute our standards to which measurements of particles of irregular shapes and differing materials must eventually be referred. A measurement of particle size then is ultimately one of effective scattering due to a particle having the diameter of a particular "equivalent" sphere used in the primary calibration.

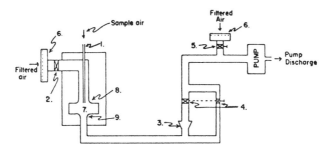

Fig. 8. Schematic diagram showing details of the air flow system for the CAES particle analyzer. 1—Sampling needle, 2—sheath-air orifice valve, 3—rotameter, 4—pushbutton valve to open rotameter circuit, 5—valve to adjust total flow, 6—filters, 7—photometer head chamber, 8—shaped inlet, and 9—shaped outlet. Note: Calibrated dilution valves may be installed on the sample-air inlet needle to change counting ranges.

Excellent discussions of sizing problems which arise as a result of physical properties of particles processed by optical analyzers are to be found in articles by Martens[14] and Whitby[15] and Vomela. Estimates to date indicate that response of our instrument to coal particles, for instance, does not differ within the precision attainable from the response to an appropriate mixture of polystyrene latex spheres having approximately the same size distribution.

ACKNOWLEDGMENTS

Development of this instrument was supported by the Coal Research Board, Department of Mines and Mineral Industries of the Commonwealth of Pennsylvania.

[1] F. T. Gucker, Jr. and C. T. O'Konski, Chem. Rev. 44, 373 (1949).
[2] C. T. O'Konski, Tech. Rep. No. 3 (1956), Office of Naval Research, Chemistry Branch, Washington, D. C.
[3] F. T. Gucker and D. G. Rose, Brit. J. Appl. Phys. Suppl. No. 3, 138 (1954).
[4] M. A. Fisher, S. Katz, A. Lieberman, and N. E. Alexander, Proc. Third Nat. Air Poll. Symposium, Pasadena, Calif., 112 (1955).
[5] A. L. Thomas, Jr., A. N. Bird, Jr., R. H. Collins III, and P. C. Rice, J. Inst. Soc. Amer. 8, 52 (July 1961).
[6] V. R. Mumma, A. L. Thomas, Jr., and R. H. Collins III, Ann. N. Y. Acad. Sci. 99, 298 (1962).
[7] D. Sinclair, J. Air Pollution Control Assoc. J. 17, 105 (1967).
[8] A. E. Martens and K. H. Fuss, Staub-Reinhalt. Luft 28, 14 (1968).
[9] W. R. Zinky, Air Pollution Control Assoc. J. 12, 578 (1962).
[10] Richard D. Cadle, Particle Size Theory and Industrial Applications (Reinhold, New York, 1965).
[11] C. N. Davies, Aerosol Science (Academic, New York, 1966).
[12] H. C. Van De Hulst, Light Scattering by Small Particles (Wiley, New York, 1957).
[13] Emanuel Parzen, Modern Probability Theory and Its Applications (Wiley, New York, 1960), p. 304.
[14] A. E. Martens, J. Air Pollution Control Assoc. 18, 661 (1968).
[15] K. T. Whitby and R. A. Vomela, Environmental Sci. Technol. 1, 801 (1967).

Airborne Particulates in New York City

Theo. J. Kneip
Merril Eisenbud
Clifford D. Strehlow
Peter C. Freudenthal

This study was undertaken to identify seasonal and source effects on the particulate contaminants of the New York City atmosphere and ultimately to relate the concentrations of these contaminants to the tissue concentrations in residents of New York City.   Continual weekly samples of particulates have been collected at three stations in the New York area on 8 by 10 in. glass fiber filters at a flow rate of 20 cfm.

The sample is ashed with a Tracerlab Low Temperature Asher and leached with nitric acid.   Metals analyzed by the Atomic Absorption method include Pb, V, Cd, Cr, Cu, Mn, Ni, and Zn.   Lead-210, total particulate, and benzene and acetone soluble organic material are also determined.

The data have been related to various meteorological parameters over a one year period to define significant seasonal and source influences, as well as site to site variations.   Very significant inverse correlations to temperature are obtained for suspended particulates, vanadium, and nickel at both Manhattan and Bronx sites.   Particulates show a less significant inverse correlation to temperature in lower Manhattan.   Oil-fired space heating sources appear to account for as much as 50% of the particulates in the Bronx at the peak demand period.

Lead, copper, and cadmium show a general inverse correlation to average wind speed, and a direct correlation to temperature.   The latter is most likely due to an inverse relation between wind speed and temperature.   The heating season input for particulates, vanadium, and nickel is so great as to overcome most of the dilution effect due to winds.   The other elements having more constant non-seasonal inputs, definitely reflect the effects of the wind.

The most significant site effect occurs with cadmium, which has a concentration in lower Manhattan three times that of the Bronx over a period of six to seven months in the summer and fall.   The differences observed for cadmium and particulates may be explained by emission source factors which have not as yet been studied.

A study was begun in 1967 to measure routinely the concentration of trace elements in the atmosphere of New York City. The ultimate purpose is to relate the concentrations of these elements to their presence in the lungs and possibly other tissues of New York residents. Analysis of lungs has in fact begun, but the data are relatively few and will not be presented here. This paper will present the evaluation of results from the analysis of samples of suspended particulates obtained by a continual sampling technique.

Jutze and Foster (1967)[1] have discussed the factors involved in developing a sampling system for ambient air monitoring of fine particulate matter. In their report they have described and recommended use of the high-volume sampler used by the National Air Sampling Network (NASN) of the National Air Pollution Control Administration, for use in obtaining 24 hr samples. They have also described the general parameters of systems satisfactory for sampling over longer time periods.

The data from the NASN for 1964 - 1965 (USPHS, 1966)[2] provide averages and frequency distributions for some thirty pollutants both gaseous and solid, including seventeen metals, from sampling stations throughout the United States. Almost no information is available on many of these substances, particularly the metals, as to seasonal differences or in many cases sampling location effects within a city.

The method of sampling described here was developed to provide weekly samples collected in preference to the daily samples of the NASN. Weekly samples offer the advantage of a seven-fold reduction in testing load, with corresponding improvement in sensitivity.

Materials or elements selected for analysis were chosen as representative of the major particulate sources and for their potential or demonstrated effects as irritants, toxicants, or carcinogens.

Experimental

Sampling System

The air sampling system for this project was designed to satisfy four requirements: (1) Of primary importance was that the flow rate of the sample be large enough and nearly constant throughout a one week sampling period (one month for a nonurban site); (2) collection efficiency had to remain nearly constant; (3) an enclosure was required to protect the filter from weather without impeding airflow; (4) the entire system had to require a minimum of attention.

These requirements were satisfied by mounting an 8 × 10 in. Gelman filter holder inside of a 20 gallon galvanized, incinerator-type refuse can with the filter surface facing upwards about 10 cm from the top. This type of refuse can has many holes along its sides which allow free passage of air through the can. The filter holder is placed just above the highest row of holes in the can, and a conventional nonperforated refuse can lid is used to cover the enclosure. The can rests on a support so that the filter is between 1.3 and 1.7 m above the building roof where the sampler is located. Air is drawn through the filter by a Roots Model 33AF blower at 20 cfm, with less than 10% variation of flow during a seven day sampling period.

The flow rate is obtained by measuring the pressure drop across a 1 in. orifice in place between the pump and the filter holder. The upstream line vacuum is also measured, although in normal operation this rarely exceeds 30 in. of water, making line vacuum corrections relatively unimportant. A bypass valve downstream of the orifice meter allows the airflow through the filter to be regulated to the desired rate.

The orifice meters were calibrated against either a Fisher-Porter rotameter or against another orifice meter which has been calibrated against the rotameter. The rotameter had been compared in the laboratory to a calibrated

Neptune dry gas meter and found to agree within one percent with the gas meter. Calibrations were made by attaching a valve and one of the calibration instruments to the sampling system orifice meter in place of the filter holder. Line vacuum and orifice pressure were recorded together with the flow rate, and a least-squares equation was fitted to the observed data.

Daily records are made of the line vacuum and orifice pressure drop and the valves are adjusted daily to restore the flow rate through the filter to 20 cfm. Under most operating conditions, the flow rate has been observed to drop less than 10% during a 24 hr period. Automatic flow control is under test and will provide a further reduction in this source of variance.

All air samples are collected on a Gelman Type A glass-fiber filter paper. The linear velocity of air across the paper is 17.5 cm sec $^{-1}$ at 20 cfm at which flow rate the paper is reported to be greater than 99.9% efficient for 0.3 micron DOP particles (Lockhart and Patterson, 1964).[3] As the filter loads it should become even more efficient.

Although it is impossible to accurately describe the airflow within the filter enclosure without involved experimentation, it is possible to make estimates to determine if there is a loss of particles by impaction onto the lid of the enclosure due to centrifugal force acting on aerosols which must pass upwards along the sides of the filter holder, and then downwards onto the filter paper. Our calculations indicate that there should be no loss of particles to sizes well above the respirable range.

Sampling Sites

The sites are indicated on Figure 1. The lower Manhattan site is near the intersection of Hudson and Houston Streets, on the roof of a twelve story building. The immediate neighborhood to the north, south, and southeast consists of similar loft buildings, housing much of the printing and lithography industry of New York. To the north-

east are the smaller residential buildings of Greenwich Village.

The 31st Street location is on the roof of a thirteen story building which is flanked to the north and south by the New York University Medical Center and Bellevue Hospital. Across the street to the east is the F. D. Roosevelt Highway and the East River. Consolidated Edison plants are located to the north on 38th and 35th Streets and to the south on 14th Street, and to the the east in Ravenswood, Queens. To the west are both old four-story residences and new high-rise apartments, and the most populous sales and offices district in New York.

On Sedgewick Avenue just north of Fordham Road in the Bronx an air sampler is operated on the roof of a three story building in a densely populated residential apartment neighborhood. The major unique sources affecting the site are an electric utilities plant across the Harlem River to the south west and the Major Deegan Expressway, which borders the back of the building.

As a background for the three locations in the city, an air sampler has been operated in a rural mountainous area of Tuxedo, New York, about 50 km northwest of New York City. Although a small apartment complex and parking lot will slightly affect this site, concentrations of pollutants observed here should generally reflect the concentrations typical of the rural northeastern United States, not greatly affected by any particular urban area.

Our data were examined in relation to meteorological variables. The data used in these studies were taken from the Central Park observations with the exception of stability data which were obtained at Kennedy Airport. The parameters which have been considered are those which affect the rates of emission (temperature), diffusion (stability and wind speed), and transport (vector wind speed). Causal relationships between pollutants and meteorological parameters may not necessarily be inferred when high correlations exist between two or more meteorological parameters. The study has not yet been extended to multivariable regression analysis or to the apportionment of the fractional variation in any measured parameter produced by each independent variable.

Analysis

The total filter is weighed before and after sampling to determine total suspended particulate matter. For trace metal and Pb-210 analysis, separate 2 × 8 in. strips of the filter (equivalent to 1000 m^3 of air) are ashed in a Tracerlab Low Temperature Asher (LTA) at a power level of 300 watts for 2 to 14 hr depending on the number of samples ashed at one time. Serious losses are believed to occur for arsenic, mercury, and possibly gallium at the power level used.

The ashed filter strips are leached twice with 40 ml of TransitAR* grade nitric acid, by heating for 1½ hr at a temperature just below the boiling point. The two nitric acid extracts, together with a distilled water rinse of the leached filter are combined and taken to near dryness. The extract is made up to 10 ml with distilled water and centrifuged for 45 min. Five milliliters of the supernatant is' diluted to 50 ml for subsequent trace metal analysis by atomic absorption. For the Pb-210 analysis, the entire extract is analyzed by the method of Petrow and Cover (1963).[4]

The filter matrix and reagent blanks for the organic extracts and Pb-210 determinations are negligible. The only metal which has an elevated blank is zinc, and this is due almost entirely to the glass-fiber filter. There has been a serious batch to batch variation in the zinc blank. One batch of filters has been found to have a high blank with serious variations between sheets, also corrosion has occurred on the stainless filters support screens introducing a nickel interference at two sites. All nickel and zinc results reported are free of these interferences.

Table I lists the detection limits and precisions for the substances tested. For the metals, the detection limit is based on the concentration necessary to give a signal twice that of background, our normal aliquot size and dilutions described above, and an average sample volume of 5000 m^3. For comparison, the average concentrations found at New York City sampling sites are listed along with typical blank values. The

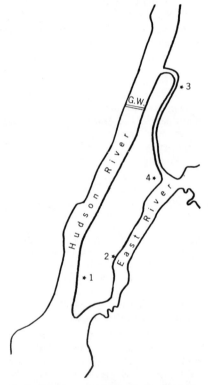

Figure 1. Air sampling sites—(1) Lower Manhattan-Hudson and Houston Streets, (2) Mid-Manhattan-31st Street and F.D.R. Drive, (3) Bronx-Sedgewick Avenue near Fordham Road, (4) NASN-121st Street and Lexington Avenue.

* TransitAR acids are available from Mallinckrodt Chemical Works, St. Louis, Mo.

Table I. Detection limits and precisions for the analysis of suspended particulates ($\mu g/m^3$).

	Det. Lim.	Blank	Av. Conc.	Av. Precision, %
Particulate	0.2	<0.2	120	2
Pb	0.028	<0.028	3.1	10
V	0.094	<0.094	1.3	21
Cd	0.001	<0.001	0.02	16
Cr	0.003	<0.003	0.1	22
Cu	0.003	0.006	0.2	9
Mn	0.002	0.003	0.06	10
Ni	0.006	<0.006	0.2	13
Zn	0.009	0.131	1.5	10
Benzene sol.	0.3	<0.3	6.1	10
Acetone sol.	0.3	<0.3	5.8	18
Pb-210[a]	0.61	<0.61	22.5	5
Ag	0.005	<0.005	<0.005	
Be	0.003	<0.003	<0.003	
Ga	0.08	<0.08	<0.08	
Hg	0.24	<0.24	<0.24	
Mo	0.06	<0.06	<0.06	

[a] Units — pCi/1000 m^3.

Table II. Annual Mean Concentrations for Airborne Particulate Contaminants 1968 Concentration ($\mu g/m^3$).

	Bronx	Lower Manhattan	Tuxedo New York	NASN[a]
Particulate	113	125	36.7	188
V	1.46	1.19	0.115	0.442
Ni	0.15	0.16[b]	0.068	0.14
Pb	3.82	2.99	0.409	1.5
Cu	0.133	0.212	0.044	0.11
Cd	0.014	0.023	0.003	
Zn	0.73[c]	2.38[d]	0.21	0.70
Mn	0.054	0.071	0.033	0.04
Cr	0.049	0.063	0.009	0.016
Pb-210[e]	22.2	23.0	21.1	
Organic				
Benzene soluble	6.0	6.2	1.6	10.9
Total[f]	10.5	13.3	3.3	

[a] E. 121st Street, Manhattan, 1964 - 65 for Particulate, 1962 - 64 for all others.
[b] Nickel data at Lower Manhattan in this and following tables is for the period up to 8/16/68.
[c] Zinc data at Bronx in this and following tables is for the period up to 6/7/68.
[d] Zinc data at Lower Manhattan in this and following tables is for the period up to 8/23/68.
[e] Pb-210 units in pCi/1000 m^3.
[f] Total organic is the sum of benzene and acetone extractables.

Table III. Annual ranges[a] for 1968 Concentrations ($\mu g/m^3$).

	Bronx	Lower Manhattan	Tuxedo New York	NASN[b] Manhattan
Particulate	68 - 149	80 - 160	24.7 - 46.7	92 - 284
V	0.60 - 2.79	0.31 - 2.38	0.04 - 0.21	0.14 - 0.87
Ni	0.06 - 0.25	0.07 - 0.21	0.01 - 0.16	0.05 - 0.25
Pb	2.54 - 5.03	1.75 - 4.15	0.11 - 0.62	0.1 - 3.0
Cu	0.03 - 0.27	0.09 - 0.30	0.012 - 0.086	0.03 - 0.25
Cd	0.006 - 0.022	0.009 - 0.036	0.001 - 0.005	0 - 0.04
Zn	0.13 - 1.44	0.83 - 4.18	0.06 - 0.26	0.01 - 1.66
Mn	0.033 - 0.081	0.035 - 0.097	0.009 - 0.057	0.01 - 0.07
Cr	0.017 - 0.093	0.027 - 0.093	0.003 - 0.023	0 - 0.035
Pb-210[c]	7.41 - 36.4	9.2 - 43.5	11.6 - 34.5	
Organic				
Benzene soluble	2.9 - 9.2	3.2 - 8.8	0.8 - 2.8	
Total	5.0 - 15.5	6.1 - 19.9	1.5 - 6.9	

a 10th to 90th percentiles.
b E. 121st Street, 1964 - 65 for particulates, 1962 - 64 for others.
c Units pCi/1000 m^3.

latter includes contributions from the filter and reagents.

The precision is given as the coefficient of variation, based on the analysis of replicate portions of a number of samples, and thus includes the variability in both uniformity of the sample across the filter and the analytical method.

Results

Annual arithmetic mean concentrations of particulates for 1968 are presented in Table II, with the reported ranges given in Table III. Data is reported for those stations which operated for the full calendar year; the roof top sites in lower Manhattan and the Bronx, and the background station in Tuxedo, New York. . As a comparison, 1964 - 1965 particulate data and 1962 - 1964 trace metal and organic soluble data from the National Air Sampling Network Site at 121st Street and Lexington Avenue are presented.

All metal concentrations observed at the lower Manhattan and Bronx sites show general agreement with the NASN with the exception of vanadium, lead, and chromium. Our lead concentra-

tions are 2 to 2.6 times higher than those of NASN. This was evidently due to losses during the muffle furnace ashing at 500°C, as reported by NASN. Volatilization of chromium and vanadium during ashing has also been reported (Dunlop, 1961).[5] Such losses account for only about 20% of the differences between our data and the NASN data for these elements. Elimination of such losses has been accomplished through the use of the LTA (Gleit and Holland, 1962).[6]

For the year 1968, there was no statistically significant difference ($P > 0.05$) for vanadium, nickel, benzene soluble, and lead-210 concentrations between the Bronx and Lower Manhattan sites. Lead was significantly higher at the Bronx site and all others were higher at the lower Manhattan site. The winter levels of particulates, vanadium, cadmium, and nickel were significantly higher than those of the summer in the Bronx, while lead, copper, and lead-210 were significantly higher in the summer. At the lower Manhattan site, only vanadium is found to be higher in the winter, while lead, cadmium, copper, and zinc were significantly higher in the

Table IV. Relationship of Meteorological Parameters.

		Correlation Coefficient			
	Temp.	Av. Wind Speed	Vector Wind Speed	Stability Index ΔT, 0700	Stability Index ΔT, 1900
Temp.	1.00	—0.92	—0.67	+0.51	+0.57
Ave. wind speed		1.00	+0.62	—0.29	—0.45
Vector wind speed			1.00	+0.02	—0.01
Stability index ΔT, 0700				1.00	+0.88
Stability index ΔT, 1900					1.00

Table V. Relationship of Pollutant Concentrations and Mean Temperature.

	Correlation Coefficient	
	Bronx	Lower Manhattan
Particulate	—0.55	—0.33
V	—0.62	—0.67
Ni	—0.68	—0.59
Pb	+0.70	+0.76
Cu	+0.74	+0.64
Cd	—0.66	+0.87
Zn	+0.11	+0.86
Mn	—0.23	—0.31
Cr	—0.002	—0.52
Pb-210	+0.45	+0.42
Benzene soluble	+0.34	+0.14

Table VI. Relationship of Pollutant Concentrations and Wind Speed.

| | Correlation Coefficient | | | |
| | Bronx | | Lower Manhattan | |
	Av. Wind	Vector Wind	Av. Wind	Vector Wind
Particulate	+0.44	+0.31	+0.32	+0.13
V	+0.58	+0.09	+0.41	+0.27
Ni	+0.51	+0.22	+0.52	+0.20
Pb	−0.78	−0.52	−0.81	−0.47
Cu	−0.64	−0.59	−0.65	−0.10
Cd	+0.76	+0.56	−0.84	−0.62
Zn	+0.16	−0.21	−0.79	−0.50
Mn	−0.01	+0.25	+0.16	+0.46
Cr	−0.03	−0.04	+0.55	+0.30
Pb-210	−0.61	−0.03	−0.58	−0.04
Benzene soluble	−0.43	−0.40	−0.23	−0.43

Table VII. Relationship of Pollutant Concentrations to Atmospheric Stability[a]

| | Correlation Coefficient | | | |
| | Bronx | | Lower Manhattan | |
	0700	1900	0700	1900
Particulate	−0.72	−0.79	−0.53	−0.55
V	−0.76	−0.89	−0.85	−0.79
Ni	−0.63	−0.74	−0.40	−0.58
Pb	−0.02	+0.30	+0.14	+0.19
Cu	+0.12	+0.18	+0.18	+0.33
Cd	−0.07	−0.20	+0.30	+0.34
Zn	−0.34	−0.46	+0.32	+0.27
Mn	−0.46	−0.37	−0.29	−0.07
Cr	+0.13	+0.11	−0.21	−0.22
Pb-210	+0.10	+0.49	+0.08	+0.42
Benzene soluble	−0.30	−0.06	−0.55	−0.42

[a] Atmospheric stability as measured by the temperature difference between the surface and a pressure level of 900 mb (normally about 950 meters). The temperature difference decreases with increasing stability, thus a negative correlation coefficient indicates that the concentration increases with increasing atmospheric stability.

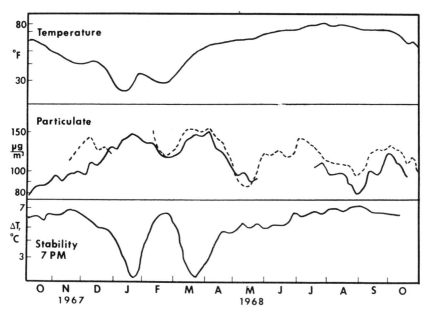

Figure 2. Four week moving average temperature, stability index, and particulate concentration, Bronx ———, Lower Manhattan ------.

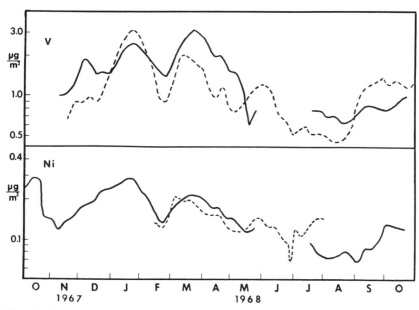

Figure 3. Four week moving average vanadium and nickel concentrations. Bronx ———, Lower Manhattan ------.

summer. Source differences most probably are involved in these seasonal and site variations.

Linear correlations between four week moving averages of meteorological parameters are presented in Table IV. All correlations have been made on this basis throughout the study to further reduce the influence of short-term variables. The stability index is the difference between the temperature at 900 mb (°C) minus the temperature at the surface measured by bidaily rawinsondes at J. F. Kennedy Airport. Although this does not describe stability immediately over the city, it does provide a measure of diffusion of pollutants outside of the urban heat bubble, as well as the potential for the bubble to form.

Linear correlations between pollutant concentrations and temperature were computed and presented in Table V. A correlation coefficient is significant at the 5% probability level in the evaluations if its absolute value exceeds 0.30 (based on the use of 50 data points) (Hoel, 1962).[7]

Average wind speed and vector windspeed correlation coefficients are given in Table VI. Fewer significant correlations are found with vector wind speed than with average wind speed. Correlation coefficients for the stability index are shown in Table VII.

The seasonal relationships of temperature, stability and particulates are shown clearly in Figure 2. The period of low temperatures accompanied by high stability results in maximum particulate concentrations. The metals which most clearly reflect this effect are vanadium and nickel as shown in Figure 3. In addition to showing inverse correlations to temperature as do the particulates, both vanadium and nickel correlate to particulates at the Bronx site, coefficients + 0.85 and +0.53; and to each other, coefficient +0.84. Particulates and vanadium are not as well correlated at the lower Manhattan site, coefficient +0.44, while the nickel particulate correlation is stronger, coefficient +0.65. The nickel-vanadium correlation coefficient is +0.83 for lower Manhattan.

During the heating season about 3.9 billion gallons of oil are consumed, some 2.6 billion gallons of it #6 grade and the remainder #2 and #4 (Heller, 1969).[8] Only 0.45 million gallons of #6 are consumed in the warm months for power generation alone, while about 0.6 million gallons are used in power generation during the cold period. Thus the use of nickel and vanadium as oil particulate tracers is fully justified by our results and the corresponding fuel use estimates.

Cadmium and copper as shown in Figure 4 demonstrate different seasonal and site variations. Copper is expected to be a major tracer element for coal burning sources. The major coal usage (90%) in New York City (Heller, 1969)[8] is at stations of the Consolidated Edison Company. These sources show little seasonal fluctuation and are large and isolated.

Lead, lead-210 and benzene soluble organic material may serve as additional tracers in particulate samples. As shown in Figure 5, lead and lead-210 are not from common sources nor are they affected in the same way by seasonal factors. Stable lead is expected to have automotive traffic as its major source while lead-210 is known to come from radon out gassing of soil (Eisenbud, 1963)[9] and might also have a significant portion from fossil fuel sources (Eisenbud and Petrow, 1964).[10]

The general behavior of lead-210 is clearly the same for both the city and rural sites. Thus both source and dispersion factors must be the same and the input from fossil fuels appears to be insignificant as compared to the natural levels.

The benzene soluble organic fraction shows a seasonal fluctuation which is similar to the results for lead and copper at the Bronx site with correlation coefficients to these elements of 0.69 and 0.50, respectively. The correlation co-

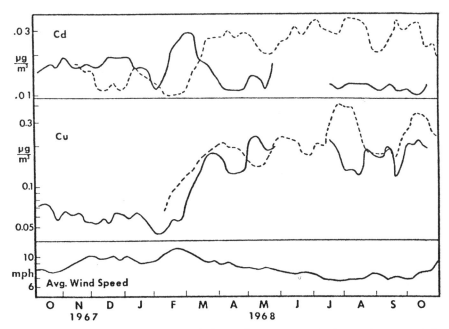

Figure 4. Four week moving average cadmium and copper concentrations and average wind-speed. Bronx ———, Lower Manhattan ·····.

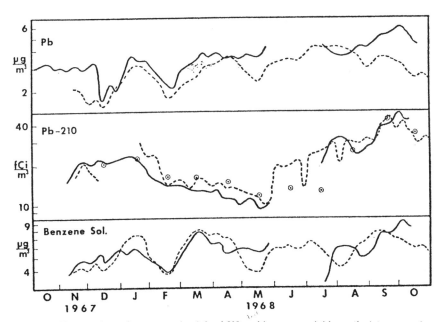

Figure 5. Four week moving average lead, lead-210 and benzene soluble particulate concentrations. Bronx ———, Lower Manhattan ·····, Tuxedo, New York ⊙⊙⊙.

efficient for benzene soluble organic-lead is much lower, 0.38, at lower Manhattan and that with copper is not significantly different from zero.

Conclusions

A number of significant correlations have been shown between the concentrations of trace elements in airborne particulates and various other factors. Among these the most important correlations are:

1. The inverse correlation of particulates, vanadium, and nickel to temperature.
2. The direct correlation of particulates, vanadium, and nickel to increasing atmospheric stability.
3. The intercorrelations of particulates, vanadium, and nickel.
4. The correlations of copper, lead, cadmium, and zinc to average wind speed.
5. The data from two sites, in lower Manhattan and West Bronx revealed the following differences

Lead	Bronx > Manhattan
Particulates	Manhattan > Bronx
Copper	Manhattan > Bronx
Cadmium	Manhattan > Bronx.
Zinc	Manhattan > Bronx
Manganese	Manhattan > Bronx
Chromium	Manhattan > Bronx
Acetone-soluble organic	Manhattan > Bronx

No differences were found for vanadium, nickel, benzene-soluble organic, and lead-210.

The correlations for particulates, vanadium and nickel conclusively demonstrate the usefulness of these two elements as tracers for particulates from oil burning sources. The elevated winter levels for these variables clearly show that the increased pollution during the winter is so great as to overcome the dispersing effects of higher winter wind speeds. For these contaminants, space heating is the principal source.

Lead, copper, cadmium, and zinc have seasonal patterns that indicate relatively constant inputs with higher concentrations related to the summer period of lower average wind speed. These elements may prove to be adequate tracers for additional sources, for example, copper is apparently related to coal burning and lead is representative of automotive pollutants.

The system developed has been used successfully to demonstrate the value of trace metals as tracers of the sources of suspended particulates. Continuing evaluation studies should provide a technique for apportioning the particulate concentrations by sources by the use of these tracers.

Acknowledgment

This investigation is supported by Contract No. C-20401 of the Department of Health, State of New York, and is part of a Center program supported by National Institute of Environmental Health Sciences, Grant No. ES-00260.

The authors wish to acknowledge the assistance of Miss Isadel Rica-Blanca and Mrs. Dorothy Wolgemuth for the work of sample preparation and analysis.

References

1. Jutze, G. A. and Foster, K. E., "Recommended standard method for atmospheric sampling of fine particulate matter by filter media—High-Volume Sampler," *J.A.P.C.A.*, **17**: 17 (1967).
2. U.S.P.H.S. *Air Quality Data from the National Air Sampling Networks and Contributing State and Local Networks 1964-1965*, U. S. Dept. of Health, Education and Welfare, Cincinnati (1966).
3. Lockhart, Jr., L. B. and Patterson, Jr., R. L., "Characteristics of air filter media used for monitoring airborne radioactivity," *NRL Report 6054*, (Mar. 20, 1964).
4. Petrow, H. G. and Cover, A., "Direct radiochemical determination of lead-210 in bone," *Anal. Chem.*, **37**: 1659 (1965).
5. Dunlop, E. C., "Decomposition and Dissolution of Samples: Organic," in Kolthoff, I. M. and Elving, P. J., eds., *Treatise on Analytical Chemistry*, Part I, Vol. 2, 1055-56, Interscience, N. Y. (1961).
6. Gleit, C. E. and Holland, W. D., "Use of electricity excited oxygen for

low temperature decomposition of organic substances," *Anal. Chem.*, **34**: 1454 (1962).

7. Hoel, P. G. *Introduction to Mathematical Statistics*, 3rd ed., 166–167, John Wiley & Sons, N. Y. (1962).

8. Heller, A. N., N. Y. City Dept. of Air Resources, private communication, (1969).

9. Eisenbud, M. *Environmental Radioactivity*, 155–160, McGraw-Hill, N. Y. (1963).

10. Eisenbud, M. and Petrow, H. G., "Radioactivity in the atmospheric effluents of power plants that use fossil fuels," *Science*, **144**: 288 (1964).

THE GRANULOMETRIC CLASSIFICATION OF AEROSOLS USING A NEW TYPE OF CASCADE IMPACTOR (FOUR STAGE AEROSOL DIFFERENTIAL SAMPLER)

L. MAMMARELLA

ABSTRACT

The importance of classification of aerosols according to diametric size is well known. In fact it is possible to find a series of differential devices working on this principle ('Cascade Impactors' of May, Orr, Lippman, 'Eolic Classificator' of Zurlo, 'Centripeter Impactor' of Hounam and Sherwood, etc.). The big interest in this field has led us to devise a four-stage aerosol differential sampler with the following principal characteristics:

(i) high efficiency and selectivity,
(ii) simplicity of use and function,
(iii) possibility of sampling high air volumes without conglomeration of particles in sampling surfaces,
(iv) employment of usual collecting surfaces (microscope slides).

The sampler gives satisfactory responses from all the above points of view.

INTRODUCTION

Many devices for collecting aerosols are based on the principles of impaction. We shall consider a gas stream to pass through a pipe at a certain speed. If we place an obstacle across the flow, the gas stream will be diverted but the particles suspended in the stream will tend to maintain their initial direction, insofar as they will settle over the surface of the obstacle.

This basic behaviour is conditioned by the following two factors: (1) the stream speed, and (2) the size of the particles. The greater the particle size, the greater the collection rate at a fixed flow speed. Conversely, the efficiency for collecting small particles can be enhanced by increasing the flow speed.

The principle of impaction sampling can be applied both to liquid and solid surface collecting devices. With regard to the latter type of device, the simplest is represented by a surface collector (e.g. a microscope slide) exposed across the jet stream or gaseous flow. Related to such a primordial system, which obviously has a low efficiency, there is a series of much more complicated, more efficient devices.

Using such instrumentation one can impact air which is convected at a known speed through a fixed hole, over a surface crossing the flow direction; the collecting surface retains the particles abandoned by the fluid stream. The collection of particles of different sizes is a primary function of the flow speed at the surface level. The system can be adapted into a series of collectors by imposing, in a single device, more impaction stages and reducing proportionally, the size of the hole for the flow passage, at each successive level. In this manner one may obtain, without varying the initial stream speed, a successive deposition of smaller and smaller particles; the number of size

100

classes will be the same as the number of stages. A four-stage cascade impactor was built in 1945 by May; this instrument greatly enlarged the performance possibilities in the air sampling field.

After May's device some others were projected and developed (the impactors of Lippman[3] and of Orr[6], the centripeter impactor of Hounam–Sherwood[2], the powder classifier of Zurlo[7], the biological impactor of Andersen[1], etc.). The efficiency of all of these impactors is good, and they are practically of the same order of accuracy. They may possibly differ in some technical and mechanical peculiarities or in their ability to sample different volumes of air. In order to increase the availability of such devices a new type of cascade impactor ('Aerosol Differential Sampler')[4,5], has been designed and developed with the following principal characteristics:

(1) simplicity and low apparatus cost,
(2) possibility of dividing a polydisperse aerosol into various size classes with a higher selectivity also for coarse aerosols,
(3) possibility of employing normal microscope slides as collecting surfaces and of assaying medium and high air volumes, without coagulation or agglomeration of particles on the collecting surfaces,
(4) settling of aerosol uniformly for making rapid and accurate evaluations.
The cascade impactor which we now describe does indeed satisfy the above mentioned requirements.

DESCRIPTION OF THE DEVICE

The four stage aerosol differential sampler (see *Figure 1*), is constructed from two substantially similar mating shells which, when fitted together, form an elongated housing. Each of these shells is subdivided into a series of chambers situated longitudinally. Closing the shells forms a series of slits which are more and more narrow from the top to the bottom. The actual slit dimensions decrease from the first to the fourth stage as follows: 1 mm–0·5 mm–0·25 mm–0·10 mm. The body is connected to a suction system. The dimensions of this device are, approximately, $16 \times 5 \times 8$ cm.

EXPERIMENTAL ASSAYS

In order to ascertain the sampling efficiency of the above mentioned cascade impactor various series of collecting controls were originally performed either employing special aerosols (fluorescent under U.V. microscope) or assaying open air. The initial employment of fluorescent particles has been considered very useful because it is possible to have a whole series of well sized powders. From a large series of such sampling assays it has been concluded also that, by comparison with other cascade impactors (essentially of the types named above), the collection efficiency is very satisfactory. This high efficiency is developed even at the first stage level (for the coarse particles). Above all it was possible to ascertain the feasibility of long period samplings (up to 1000-1500 litres of air) without producing a high degree of particle conglomeration on the collecting surfaces. Obviously the volume of air treated is a function of the pollution level. This is primarily because of the amplitude of the collecting strip (about 6 centimeters); in this way the sampling becomes statistically valid.

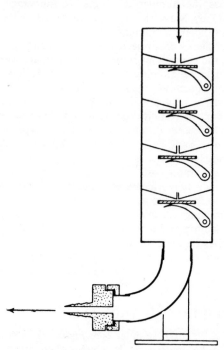

Figure 1. Four stage aerosol differential sampler.

Figure 2 shows a comparison of collection between an industrial area (left), an urban area (middle) and a suburban area (right), at the same time on a winter's day. The collection volume was 300 litres of air at a speed of 30 litres/minute.

SOME ADVICE FOR AN APPROPRIATE SAMPLING

First of all, it is necessary to clean carefully the interior of the collector before starting. This operation is particularly easy because the device may be completely opened. The separation between the slit and the corresponding microscope slide is assured by two guides engraved in the body of the collector. The differences between the slits and the collecting surfaces are strictly proportional to the breadth of each slit. Before use, the slide must be clean but great care is not necessary; an inexperienced operator need be only briefly trained to observe the aerosol background in the slide under a microscope. Very rapidly he will be able to differentiate this background from the depositing strip. In fact the latter is very characteristic and very regular.

SUMMARY

As a result of the various experiments made, the selective efficiency of the various stages of an aerosol differential sampler was determined. In this context it is worth noting that the selectivity of the impactor may vary, the

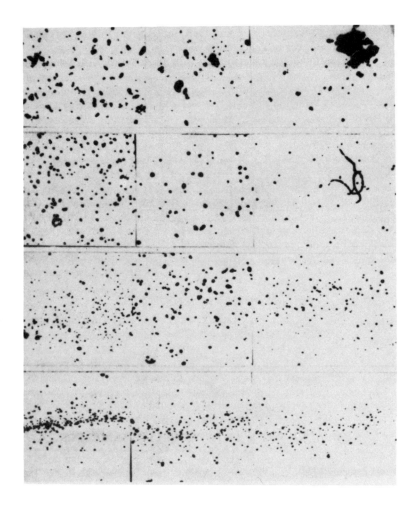

Figure 2. Comparison of aerosols collected from three different areas *(left:* industrial, *centre:* urban, *right:* suburban).

dimensional values both diminishing the breadth of slits (operationally not easy), and varying the suction speed. One may claim that the apparatus has a good selectivity at a reduced suction speed (20-30 litres/minute) especially when aerosols are grouped in size classes of multiples of 5 microns.

The employment of the linear cascade impactor that we have described is as simple as the methods of counting particles. This is principally due to the following:

(1) The very regular deposition of aerosols over the collecting strip; consequently it is only necessary to count over a few microscope fields

103

randomly chosen along the strip and then multiply the number of particles per field by the number of fields in the whole strip.

(2)   The situation of the collecting strip along the middle of the microscope slide.

(3)   The noticeably large total area of collecting surfaces, which allows the evaluation of medium and high air volumes and, therefore, gives statistically valid readings.

References

[1] A. A. Andersen. *J. Bacteriol.* **76**, 471–484, (27 Nov. 1958).
[2] Hounam–Sherwood. cf. ref.[5]
[3] M. Lippman. *Amer. Ind. Hygiene Assoc. J.* **22**, 3 (1961).
[4] L. Mammarella and L. di Giambernardino. Prove di prelievo comparativo fra il 'cascade impactor' (MAY) e il 'campionatore differeniale a 4 Stadi (Mammarell 1° modello).
[5] L. Mammarella. Inquinamenti dell'aria e loro rilevamento—"Il Pensiero Scientifico Editore"–Roma–nov. 1968.
[6] C. Orr, Jr. Cascade Impactor for Sampling Smokes, Dust, Fumes—U.S. Patent Office—2–967–164—August 2 1960.
[7] N. Zurlo. Nouveau pneumo-classificateur pour poussiéres fines–Comm–au "Quatriénne colloque sur les poussiéres" Institut Nationel sur le securite. 28–30 III, 1957, Paris.

Particle Sampling Bias Introduced By Anisokinetic Sampling and Deposition Within the Sampling Line

G. A. SEHMEL, Ph.D.

Particle sampling errors from several sources were studied with monodisperse aerosol particles. The sources of errors investigated were the influence of the inlet probe diameter on isokinetic samples, the turbulent deposition of particles in straight tubes and one-elbow configuration, and non-isokinetic flow through the sampling probe. The sampling errors were shown to be significant for particles as small as 1 micron in some cases. A particularly significant observation was that non-uniform concentration in the conduit sampled can cause serious errors. Some studies of the particle concentration profiles in a three-inch tube showed that particles in many cases are concentrated in a narrow annulus near the wall, making representative sampling nearly impossible for these cases. The research emphasizes the need for a thorough knowledge of many particle interactions with the air system to be sampled and with the sampling arrangement chosen.

Introduction

PARTICLE SAMPLING ERRORS must be predicted when sampling airborne particles with a cylindrical sampling probe. These errors result in large measure from the deviation of particle motion from air motion. Mathematical models and experimental data have been obtained to predict these errors. However, in general, satisfactory correlations have not been developed to predict the sampling errors.

The purpose of the present study was to determine errors in sampling an aerosol being carried in turbulent flow in a relatively small diameter tube. The errors are those resulting from using sampling probes facing into the air flowing in a tube and for particle deposition in the sample delivery line. Sources of error considered are: (a) concentration pro-

This paper is based on work performed under United States Atomic Energy Commission, Contract AT(45-1)-1830.

files—the failure of one sample location to adequately represent the average particle concentration in a sampled tube; (b) sampling—anisokinetic sampling and the influence of probe diameter on isokinetic sampling; and (c) deposition—particle deposition in curved sampling probes and in straight sample delivery lines.

Literature Background

These three sources of sampling error will be discussed separately within each of the following sections.

Profiles

The particles flowing through a tube are usually sampled to determine the average particle concentration. However, accurately sampling at one location within the tube will not yield the average concentration if radial concentration gradients are present.

Radial concentrations of particles within the tube can be caused by the forces of diffusion, electrical charge effects,[1] centrifugal forces due to air swirl, and re-entrainment forces. In most cases these profiles can not be predicted.

Sampling

The sampling errors for anisokinetic sampling have been determined for several systems. As expected, the results showed that for subisokinetic sampling the apparent concentrations are greater than that of the sampled airstream. Similarly for greater than isokinetic sampling, the apparent concentrations are too low.

The concentration ratio of the apparent to true particle concentration should be predictable from mathematical modeling of sampling. The dimensionless equations of motions for the models indicate that the concentration ratio should be a function of the Stokes number

$$STK = S/D \qquad (1)$$

where S is the particle stopping distance and D is the sampling probe diameter.

The Stokes number was used by Lundgren and Calvert[2] to correlate their data for sampling into side port probes. Similarly, the Stokes number was used by Sehmel[3] to correlate his data for sampling into a filter held in the air stream.

For sample probes aligned with the air velocity, a theoretical study of sampling into probes has been reported by Vitols[4] who obtained agreement between predictions and the data of Badzioch[5] and Hemeon and Haines.[6] In these experimental studies, polydispersed particles were used. In comparison, Watson[7] used 4 and 32 μm spores for anisokinetic sampling.

For sample probes non-aligned with the air velocity Mayhood and Langstroth, as reported by Watson,[7] showed a decreased collection of 4, 12, and 37 μm MMD diethyl phthalate particles for isokinetic velocities. The apparent concentration decreased with an increase in the angle between the sample probe inlet and the air velocity and also an increase in particle size.

The apparent concentration is also a function of the sampling probe diameter. For isokinetic velocities, Griffith and Jones used 300 μm MMD coal dust and Mayhood and Langstroth used 37 μm MMD diethyl phthalate, as reported by Watson,[7] to show that the apparent concentration was equal to the true concentration for probe diameters from about 0.6 to 3.6 cm. However, the apparent concentration decreased to 0.6 of the true concentration for a probe diameter decrease to 0.4 cm ID.

Deposition

Sampling errors also result from particle deposition in the sample delivery line downstream of the sample probe inlet. Although little research has been reported on deposition in tube bends,[8] both mathematical models[9-12] and experimental[9,13-16] data have been reported in depth for turbulent deposition in straight tubes.

Particle deposition from turbulent flow is usually reported as a deposition velocity equal to

$$K = \frac{\text{NUMBER OF PARTICLES DEPOSITED/UNIT AREA/UNIT TIME}}{\text{NUMBER OF PARTICLES/UNIT VOLUME OF BULK FLOW}}$$

$$\tag{2}$$

which has the units of length/time. This deposition velocity is used to calculate the decrease in bulk average particle concentration, C_0, across the tube as a function of tube length:

$$\ln \frac{C_{0_2}}{C_{0_1}} = -4 \frac{K}{V} \frac{L}{D}, \tag{3}$$

where C_{02} is the average concentration at a distance, L, downstream of a point where the average concentration is C_{01}; V is the average air velocity, and D is the tube inside diameter.

Models to describe the turbulent deposition of particles have been proposed by Friedlander and Johnstone,[9] Owens,[10] Davies,[11] and Wells and Chamberlain.[12] These models are based upon the assumptions that deposition surfaces are perfect particle sinks and that deposition is by a two step process. The first process is by eddy diffusion from a turbulent region of uniform concentration. The diffusion process terminates at a stopping distance from the surface. This stopping distance, S_s, is the distance that a particle with a velocity equal to the diffusion velocity would travel in still air. The particle is brought to a stop by air drag which opposes the momentum of the particles.

Disagreement exists in the assumptions as to the proper value for this diffusion velocity at the start of the stopping distance. In order to resolve the disagreement, Sehmel[17] used a diffusion model to calculate the required diffusion velocities which would yield agreement with available experimental deposition data. The calculated diffusion velocities showed a wide range of values which included those assumed for the several deposition models. The author concluded that current models do not adequately explain turbulent deposition. Consequently, Sehmel developed an eddy diffusion correlation for the diffusion of particles approaching the perfect sink surfaces of vertical tubes. These results were calculated from experimental deposition velocity data and show that the effective eddy diffusivity of particles is greater than the eddy diffusivity of air momentum.

Experimental

Single sized particles produced with a spinning disc aerosol generator were used to study radial concentration profiles in tubes, errors resulting from certain sampling flows and probe configurations, and deposition in tubes. The particles were either uranine or a uranine-methylene blue mixture which had a density[18] of 1.5 g/cm^3.

For each experiment the particles were generated, electrically neutralized,[8] passed into a holding chamber, and then the particle-laden air was withdrawn upwards through vertical tubes which were the test sections. The test sections were up to 50 feet in length and had inside diameters of 0.21, 0.62, 1.152, and 2.81 inches. Each type of experiment will be discussed separately.

Profiles

For determining the radial concentration profiles in the test sections, a filter was placed across the test section within a slip joint in the tubing. After particle collection, the filter was removed and cut into concentric annuli. These filter sections were washed with distilled water and the wash liquid analyzed fluorimetrically to determine the mass of uranine collected. The mass flux through each annulus was then normalized to the flux passing through the filter center.

Sampling

For the sampling measurements, a vertical aluminum tube with 2.81 inches ID served as a small diameter wind-tunnel. This tunnel was sampled at its axis through curved sampling probes with inlets ground to knife edges.

Sampling velocities were both isokinetic and anisokinetic. For the anisokinetic velocities, the sampling probe was an aluminum tube with 0.402 inch ID and 0.049 inch wall. This probe was formed to a 4.4 inches radius of curvature with an arc of about 60 degrees. For the isokinetic velocities, probes of several diameters were used of nearly the same over-all length.

The sample probes were held firmly in place. The probes passed through saddle pieces to which they were welded. These probes were inserted into a mating hole in the 2.81 inches ID tunnel. When the gasketed saddle was clamped into place, the centerline of the probe inlet coincided with the tunnel axis.

The experimental procedure for determining sampling errors was as follows: A sample was withdrawn through the probes at a controlled constant flow rate and the particles were subsequently collected on a glass-fiber filter. The filter and probe were removed. The quantities of uranine deposited inside the probe and on the filter were determined through dissolving the uranine and measuring the fluorescence of these solutions with a calibrated fluorimeter. Some particles would deposit on the outside of the probe and the downstream walls of the tunnel. All remaining particles were collected on a glass-fiber filter through which the total air flow passed. The downstream tube surfaces and the filter were washed to dissolve the deposited uranine. Uranine in the solutions was measured fluorimetrically and a mass balance determined for particles in the tunnel.

Deposition

Turbulent deposition was measured by drawing single sized particles from a holding chamber through vertical tubes. Particles not deposited were collected on a filter at the end of the tube.

After a run the tubes were cut into short lengths in order to determine deposition as a function of tube lengths. Particle mass balances were made by washing the tube lengths and filter and subsequently analyzing these solutions fluorimetrically.

Results and Discussion

The experimental data will show that particles may be non-uniformly distributed across a tube, and that the non-uniformity is a function of particle size, flow rate, and tube surface. This non-uniformity compromises the validity of samples taken from tubes. Also, sample validity can be compromised by deposition in the sampling probe and sampling line.

Profiles

The filter loadings were normalized to the filter loading at the filter center. These filter loadings are shown in Figures 1 and 2 as a function of the arithmetic average reduced radius for each annular filter section.

Filter loading profiles shown in Figure 1 are for particle diameters of 6, 12, and 28 μm for a constant Reynolds number of 55,000 in a 2.8 inches ID tube. These profiles are a function of both particle size and the nature of the tube surfaces. For a tube surface made tacky with petroleum jelly, the filter loading decreases as the tube wall is approached. This type of profile has been found for all ranges of variables studied for tacky surfaced tubes. In contrast, the other data shown in Figure 1 are for a dry untreated tube. The data for 28 μm particles show that the normalized filter loadings adjacent to the tube wall can be greater than unity. This type of profile is termed an "inverse" profile.

The filter loading profiles are independent of the experimental range of pressure drop across the filter. For 12 μm particles, the profile is the same for pressure drops of 3 and 78 inches of water. The filter with the lower pressure drop was used in obtaining data shown in Figure 2.

The filter loading profiles are a function of the Reynolds number. In Figure 2 for dry surfaced tubes, the filter loading profiles are shown for 28 μm particles as a function of Reynolds number from 647 to 55,600. The filter loading adjacent to the wall is seen to vary by a factor of four $(1.6/0.4 = 4)$. The data obtained for the untreated tube wall cases considered do not fit a pattern which allows extrapolation or prediction. For the present the data for the phenomenon must be regarded as empirical.

111

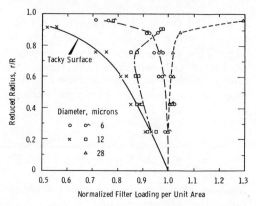

FIGURE 1. Particle loading profiles at a Reynolds number of 55,000 (98 ft³/min) in a 2.81 inches ID vertical tube (filter pressure drop, inches H_2O: $O = 3$, $\sigma = 78$).

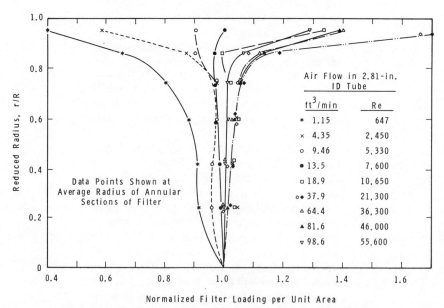

FIGURE 2. Filter loading profiles for $28\mu m$ diameter uranine particles.

Concentrations determined from these filter loadings would be equal to the true airborne particle concentrations if the effects of the changes in air velocity profiles on the particle motion could be taken into account. This has not been done due to the uncertainties of particle inertia and turbulent flow on particle motion. However, the implications of non-uniform filter loadings are important to sampling particles from a transporting conduit since particle concentrations are proportional to the filter loading.

The conclusion from the non-uniform filter loadings is that significant concentration gradients can exist in a tube of the size considered when turbulent flow exists in the tube. The concentration profiles are further substantiated by the sampling studies.

Sampling

Both isokinetic and anisokinetic sampling results are reported as concentration ratios. This concentration ratio is defined as the ratio of the observed sample concentration to the bulk average concentration in the tunnel. Concentrations are calculated from mass balances of particles and air passing a given cross section of the tunnel at the level of the sample probe inlet. The bulk average particle concentration is the ratio,

$$C_0 = \frac{P}{Q}, \tag{4}$$

where P is the total number of particles and Q is the total volume of air. These quantities are defined as

$$P = P_S + P_R, \tag{5}$$

and

$$Q = Q_S + Q_R, \tag{6}$$

where the subscripts S and R refer to respectively the amounts entering the sampling probe and remaining in the wind tunnel. Similarly, the observed concentration,

$$C = P_S / Q_S, \tag{7}$$

is that obtained through the sample probe.

If sampling from a point in the cross section of the tube were truly isokinetic, this ratio would reflect only the particle concen-

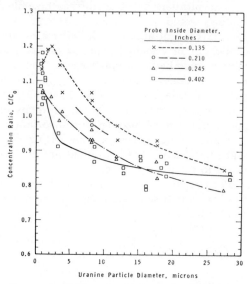

FIGURE 3. Effect of sample probe diameter on concentration ratios for isokinetic velocity based on the maximum flow velocity of 18.0 to 19.6 ft/sec in the tunnel (tunnel Re 26,400 to 28,700).

tration profile. In practice, the concentration ratio also reflects the effects of probe diameter and anisokinetic sampling.

A. Isokinetic Sampling—Effects of Probe Diameter

Isokinetic sampling velocities were made equal to the maximum[19] centerline velocity for turbulent flow in the tunnel. From the samples, the concentration ratios, C/C_0 were calculated and are shown in Figure 3 for four probe diameters as a function of particle size for a constant air velocity.

These concentration ratios for isokinetic velocities are near unity and range from about 0.8 to 1.2. These ratios are a function of both probe diameter and particle size.

Much of the decrease in concentration ratio for particle sizes from 3 to 28 μm can be attributed to differences in particle concentration profiles as a function of particle size. It was shown in Figure 1 that the centerline concentration (filter loading) decreases with

114

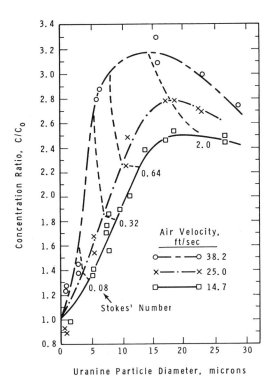

FIGURE 4. Effect of air velocity upon concentration ratios when sampling at 0.3 of the isokinetic velocity based on the average velocity in the tunnel (0.40 inch ID Probe).

respect to the average concentration across the tube as the particle size is increased from 6 to 28 μm. The relation between concentration ratios and profiles will be illustrated for one particle size. For 28 μm particles, the concentration ratios range from 0.78 to 0.85. Isokinetic concentration ratios less than unity are indicative that the particle concentration at the tunnel axis must be less than the bulk average concentration across the tunnel. This shape of concentration profile is confirmed by the filter loadings in Figure 2. For the same Reynolds number, the filter loadings show that the concentration profile is inverse. About 1.4 to 1.6 times as many 28 μm particles per unit area are found near the filter perimeter than the center of the filter.

The apparent anomaly of the isokinetic sample drawn from near the axis of the conduit

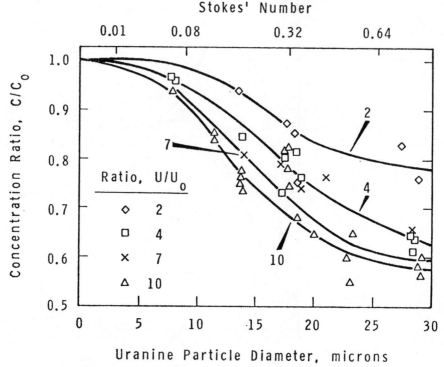

FIGURE 5. Effect of above isokinetic sampling velocity based on a constant average velocity of 3.66 ft/sec in the tunnel (0.402 inch ID probe).

showing as much as 20% higher than the bulk concentration has not been satisfactorily explained. The effect is much greater than could be attributed to experimental errors in setting flow rates and in determining particle mass balances. The effect may be caused by changes in air flow velocities and air turbulence resulting from the presence of the sample probe in the tunnel. The effect is more pronounced for smaller diameter sampling probes (and particles), yet is apparent for the 0.402 in. diameter probe, as well.

B. Anisokinetic Sampling

Errors resulting from subisokinetic and above isokinetic sampling velocities were studied. Since anisokinetic velocities distort the air velocity profiles in the tunnel, the isokinetic velocity has been assumed to be equal to the average velocity in the tunnel.

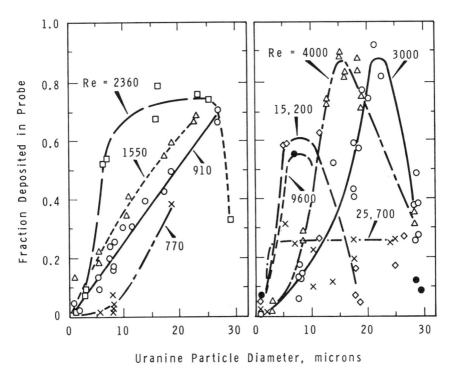

FIGURE 6. Deposition within a 4.7 inches long curved sampling probe (at Re = 2,000, flow is 0.51 cfm in the 0.402 inch ID probe).

Concentration ratios for sampling at 0.3 of the average velocity are shown in Figure 4 as a function of particle size and average velocity in the wind tunnel. These concentration ratios are greater than unity and increase rapidly with particle size. A most significant result is that the concentration ratio is from 1.4 to 2.5 for particles as small as 5 μm.

The increase with particle size is expected since the approaching air tends to be diverted around the sample probe inlet while particle inertia tends to cause particles to be driven into the probe. Thus, more particles are collected in the sample probe than were originally in the air which enters the probe. It follows that the sample concentration should increase with particle size.

For sufficiently large particles, the particle inertia would cause the concentration ratio to

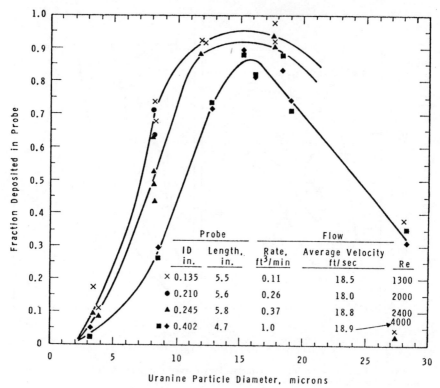

The figure contains the following legend table:

	Probe		Flow		
ID in.	Length, in.	Rate, ft³/min	Average Velocity ft/sec		Re
× 0.135	5.5	0.11	18.5		1300
● 0.210	5.6	0.26	18.0		2000
▲ 0.245	5.8	0.37	18.8		2400
■ ◆ 0.402	4.7	1.0	18.9		4000

Y-axis: Fraction Deposited in Probe

X-axis: Uranine Particle Diameter, microns

FIGURE 7. Deposition within curved sampling probes as a function of probe ID (probes are curved to a 4.4-inch radius of curvature).

approach 3.33 if the particles were uniformly distributed across the tunnel. However, the observed concentration ratios show maxima below 3.33. These maxima result from the "inverse" profiles for the larger diameter particles. For these particles relatively more particles are at the tube walls than at the tube center.

Also shown in Figure 4 are the Stokes number for each case. As stated previously, the Stokes number has been used to correlate observed concentration ratios. However, the conclusion from the present study is that concentration ratios are not a simple function of Stokes number.

Similar results are shown in Figure 5 for sampling velocities from 2 to 10 times the average velocity in the tunnel. The concentration ratios decrease below unity as the particle size is increased. For these flow conditions, some of the particles originally in the air which enters the sample probe will not be drawn into the probe because of particle inertia. Thus, the sample concentration should decrease as particle size is increased.

C. Deposition in Sampling Probes

The importance of deposition in the curved sampling probe was also determined for the sampling conditions. Deposition is shown in Figure 6 as the fraction of material entering the probe which is deposited in the 4.7-inch long probe. The deposition can be significant and is a function of both particle size and Reynolds number based on the average flow rate. For any Reynolds number, the fraction deposited increases rapidly until a maximum deposition range occurs. The leveling off or decrease in deposition for larger sized particles is considered to be caused by effects of particle re-entrainment.

Deposition is a function of laminar and turbulent flow. Within the Reynolds number transition range from 2360 to 3000, the deposition mechanisms[20] appear to have changed. For Reynolds numbers below 2360 and for Reynolds numbers above 3000 the deposition for each Reynolds number range increases with Reynolds number until re-entrainment effects become significant. The difference between the two ranges is that the deposition of 20 μm or smaller particles is greater for the lower Reynolds number of 2360 than for the higher Reynolds number of 3000.

In addition to Reynolds number, the effect of probe size on probe deposition was obtained from the sampling studies. The deposition in the four curved sampling probes are compared in Figure 7 for an average velocity of 18.0 to 18.9 ft/sec. These results show that probe deposition increases for a decrease in probe diameter for a constant flow rate and particle size.

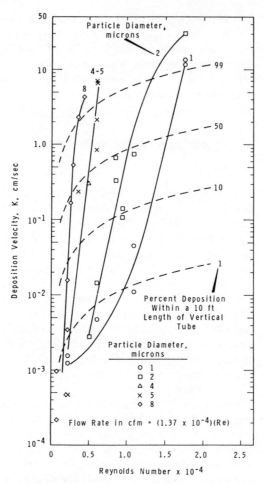

FIGURE 8. Deposition as a function of Reynolds number of methylene blue-uranine particles in 0.21 inch ID vertical tubes.

Turbulent Deposition in Straight Tubes

The deposition of particles was measured for turbulent flow through vertical tubes. The data reported here are for fully developed flow and do not include the effects of flow disturbances on deposition.[16] The deposition results are shown as deposition velocities, K, in Figures 8 to 11.

In Figure 8 the deposition velocities to a dry surfaced tube of 0.21 inch ID are shown as a function of the Reynolds number with particle size as a parameter. The data points are represented by solid curves which show that deposition velocities increase rapidly with both flow rate and particle size. The solid curves are intercepted by broken-line curves which show the calculated percentage deposition within a 10-foot length of vertical tubing.

In Figure 9 deposition velocities are shown for a larger tube of 0.62 inch ID. Similar trends are shown as for the smaller diameter tube. However, the deposition velocities tend to increase more slowly with an increase in Reynolds numbers when the deposition velocities are of the order of 10 cm/sec. This leveling off is attributed to effective particle re-entrainment from the tube surfaces caused by particles not being held to the tube surface.

In Figure 10, deposition velocities are shown for a larger tube of 1.152 inch ID. For this tube size, the range of variables was extended to more clearly show effective re-entrainment conditions for dry walled tubes. For the 8 to 9 μm particles, deposition increases with flow rate up to a maximum at a Reynolds number of about 17,000. For an increase in Reynolds numbers to 21,000 the deposition velocity decreases which again shows effective particle re-entrainment.

Re-entrainment is a function of the stickiness of the tube wall. This is shown in Figure 10 for 8 to 9 μm particles for Reynolds numbers greater than 21,000. The 8 to 9 μm curve branches into two curves at a Reynolds number of 21,000. The upper curve is for tubes covered with a thin layer of petroleum jelly to form a tacky coated tube. This surface acts as a perfect particle sink. In contrast, the lower curve is for untreated or dry walled tube surfaces. It follows that particles must be re-entrained from the untreated tube surface for Reynolds numbers greater than 21,000.

The deposition velocities for re-entrainment Reynolds numbers in untreated tubes are the observed total effects of many variables. These effects include tube surface roughness,

121

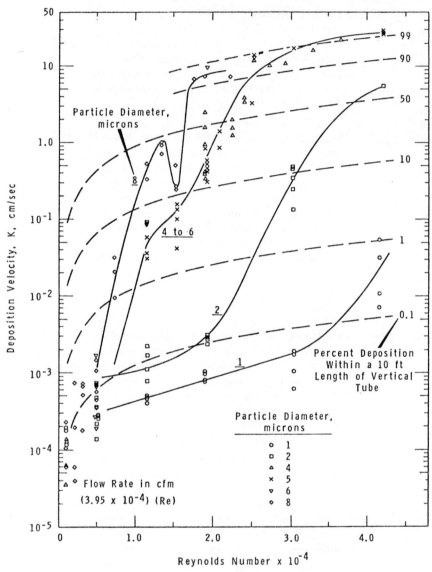

FIGURE 9. Deposition as a function of Reynolds number for methylene blue-uranine particles in 0.62 inch ID vertical tubes.

electrical effects from equilibrium charges on the particles, and any residual stickiness effects due to either the particle or the tube surface. The total effects of these variables is not always the same. This is concluded for the low-

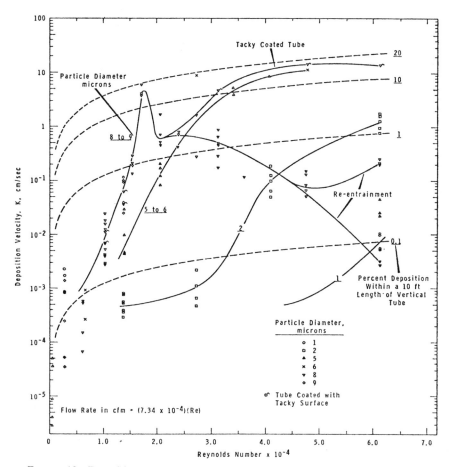

FIGURE 10. Deposition as a function of Reynolds number for methylene blue-uranine particles in 1.152 inches ID vertical tubes.

er branch of the 8 to 9 μm curve. This curve shows a second branch at a Reynolds number of 45,000. The deposition velocities show a very wide separation at a Reynolds number of 61,000.

The deposition velocities for particle sizes up to 29 μm in size are shown in Figure 11 for untreated 2.81 inch ID tubes. In contrast to the data for smaller size tubes, the deposition velocities are shown here as a function of particle size with Reynolds number as the parameter. The data trends are similar to those for the smaller sized tubes.

The deposition velocities shown in Figures 8 to 11 can not be accurately predicted.[15,21,22] Consequently, the data are represented by smoothed curves. Theories to predict deposition have been critically reviewed by Sehmel[17] who determined that predictions from existing theories did not show statistical agreement with experimental results. Consequently, a model was developed in which the deposition processes are described by effective particle eddy diffusion coefficients which are greater than the eddy diffusion coefficients of air momentum. Deposition velocities predicted from these effective particle eddy diffusion coefficients show statistical agreement with experimental deposition velocities. The predictions are correlated with the experimental data, but the correlation shows a large variance.

Linear regression was also used to develop empirical correlations to represent the dimensionless deposition velocity

$$K^+ = \frac{K}{u_*} \qquad (8)$$

where the friction velocity is

$$u_* = V\sqrt{f/2} \qquad (9)$$

and the Fanning friction factor[19] is

$$f = 0.00140 + 0.125 (Re)^{-0.32} \qquad (10)$$

Correlations were developed for the cases in which the vertical tube surfaces were a perfect particle sink and also for the case for which particles might be re-entrained after deposition. The correlation to perfect sink surfaces in vertical tubes is

$$K^+_{pred} = 1.47 \times 10^{-16} \rho^{1.01} R^{2.10} Re^{3.02},$$

where ρ is the particle density in g/cm^3, R is the ratio of particle diameter in microns to tube diameter in cm, and Re is the Reynolds number for flow in the tube.

124

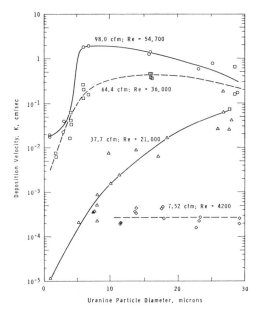

FIGURE 11. Deposition as a function of particle diameter for methylene blue-uranine particles in 0.281 inch ID vertical tubes.

All tubes, however, are not necessarily perfect particle sinks. The correlation for all the deposition data for both non-perfect and perfect sink tubes is

$$K^+_{pred} = 1.0 \times 10^{-17} \, \rho^{1.83} R^{2.99} Re^{3.08}.$$

Experimentally it has been shown that deposition to perfect sink surfaces is always equal to or greater than deposition to non-perfect sink surfaces. However, predictions from equations 11 and 12 do not always show the greater deposition. This is a result of the data scatter and the fact that these equations are simple correlation. Particle penetration predicted from these equations are shown in Table I for 200, 500, and 2000 cm lengths of vertical tubing. The predicted fraction penetrating these lengths are from 0.0 to 1.00. Consequently, this table is presented to indicate a first approximate of deposition for many practical conditions.

125

TABLE I

Fraction of Entering Particles which Will Not Deposit within Lengths of Vertical Tubes as Predicted from Equation (11) for Perfect Sink Surfaces

[In brackets, equation (12) for all data]

d_p μm	Tube Diameter cm	Tube Reynolds Number	Flow Rate cc/sec	ρ = 4 — 200	ρ = 4 — 500	ρ = 4 — 2000	ρ = 6 — 200	ρ = 6 — 500	ρ = 6 — 2000	ρ = 8 — 200	ρ = 8 — 500	ρ = 8 — 2000
1	0.5	4,000	233	0.978 (0.986)	0.946 (0.965)	0.802 (0.867)	0.967 (0.971)	0.920 (0.928)	0.717 (0.741)	0.957 (0.951)	0.895 (0.881)	0.641 (0.602)
	1.0	6,000	699	0.992 (0.997)	0.980 (0.993)	0.921 (0.971)	0.988 (0.994)	0.969 (0.985)	0.883 (0.940)	0.983 (0.990)	0.959 (0.974)	0.846 (0.900)
	2.0	8,000	1864	0.998 (1.000)	0.994 (0.999)	0.978 (0.996)	0.997 (0.999)	0.992 (0.998)	0.967 (0.991)	0.996 (0.998)	0.989 (0.996)	0.956 (0.985)
	4.0	10,000	4660	1.000 (1.000)	0.999 (1.000)	0.995 (0.999)	0.999 (1.000)	0.998 (1.000)	0.993 (0.999)	0.999 (1.000)	0.998 (1.000)	0.990 (0.998)
2	0.5	4,000	233	0.910 (0.893)	0.789 (0.753)	0.388 (0.322)	0.867 (0.788)	0.700 (0.552)	0.241 (0.093)	0.827 (0.669)	0.621 (0.365)	0.149 (0.018)
	1.0	6,000	699	0.965 (0.977)	0.915 (0.943)	0.701 (0.790)	0.948 (0.952)	0.875 (0.884)	0.586 (0.610)	0.931 (0.920)	0.836 (0.811)	0.489 (0.434)
	2.0	8,000	1864	0.991 (0.997)	0.977 (0.991)	0.909 (0.966)	0.986 (0.993)	0.965 (0.982)	0.867 (0.930)	0.981 (0.988)	0.953 (0.970)	0.826 (0.884)
	4.0	10,000	4660	0.998 (1.000)	0.995 (0.999)	0.979 (0.996)	0.997 (0.999)	0.992 (0.998)	0.969 (0.991)	0.996 (0.999)	0.989 (0.996)	0.958 (0.985)
6	0.5	4,000	233	0.387 (0.049)	0.093 (0.001)	0.000 (0.000)	0.239 (0.002)	0.028 (0.000)	0.000 (0.000)	0.148 (0.000)	0.008 (0.000)	0.000 (0.000)
	1.0	6,000	699	0.700 (0.401)	0.410 (0.208)	0.028 (0.028)	0.584 (0.037)	0.261 (0.037)	0.005 (0.000)	0.488 (0.005)	0.166 (0.004)	0.001 (0.000)
	2.0	8,000	1864	0.909 (0.912)	0.788 (0.794)	0.385 (0.398)	0.866 (0.824)	0.698 (0.616)	0.238 (0.144)	0.825 (0.720)	0.619 (0.441)	0.146 (0.038)
	4.0	10,000	4660	0.979 (0.989)	0.948 (0.972)	0.809 (0.894)	0.969 (0.977)	0.923 (0.943)	0.727 (0.790)	0.958 (0.961)	0.899 (0.905)	0.653 (0.672)
10	0.5	4,000	233	0.062 (0.000)	0.001 (0.000)	0.000 (0.000)	0.015 (0.000)	0.000 (0.000)	0.000 (0.000)	0.004 (0.000)	0.000 (0.000)	0.000 (0.000)
	1.0	6,000	699	0.353 (0.055)	0.074 (0.000)	0.000 (0.000)	0.208 (0.002)	0.020 (0.000)	0.000 (0.000)	0.122 (0.000)	0.005 (0.000)	0.000 (0.000)
	2.0	8,000	1864	0.757 (0.654)	0.498 (0.346)	0.061 (0.014)	0.657 (0.410)	0.350 (0.108)	0.015 (0.000)	0.570 (0.221)	0.246 (0.023)	0.004 (0.000)
	4.0	10,000	4660	0.940 (0.950)	0.857 (0.879)	0.538 (0.597)	0.911 (0.897)	0.792 (0.763)	0.393 (0.338)	0.883 (0.832)	0.732 (0.632)	0.287 (0.160)

Summary and Conclusions

Several sources of error in sampling and interpreting sampling data have been discussed which could be important. The following conclusions were reached for the range of variables studied:

1. Although isokinetic sampling is generally accepted as an accurate method, this study showed that for sampling from a 2.81 inch ID duct the sampling probe diameter may affect the accuracy of sampling even when sampling flow is isokinetic. Small diameter inlet probes give up to 20% bias to the measured concentration.

2. For subisokinetic sampling, the concentration ratio increases rapidly with particle size. The concentration ratio was from 1.4 to 2.5 for 5 μm particles. This means that subisokinetic sampling errors must be considered even for particles smaller than 5 μm diameter.[23]

3. For all anisokinetic sampling velocities, the concentration ratios were not simply correlated with the Stokes number.

4. Particle deposition in straight and curved tubes may be significant. However, the deposition data in curved tubes has as yet not been adequately correlated to predict deposition for all situations.

5. An additional important source of sampling error will arise from non-uniformity of particle concentration in the cross section sampled. For the small diameter particle-transporting duct studied, a tacky-walled tube gave fairly uniform concentrations, but an untreated tube gave a depleted region in the central part of the tube and an enhancement in the region near the wall. This effect was most prominent for the largest particles studied, 29 μm. These sources of error must be anticipated in sampling carried out under similar conditions.

In summary, small diameter probes should be avoided to insure an accurate sample. The

127

sample should be as short as possible and should not include tube bends. If these conditions are compromised, the results of the present study can be used to estimate the resulting sampling errors.

Nomenclature

C = Number concentration of particles per unit air volume at tunnel axis that is indicated by the collected sample, P_S/Q_S.

C_o = Bulk average number concentration of particles per unit air volume across the tunnel at the level of the sample probe inlet, P/Q.

d = Particle diameter.

D = Tube inside diameter.

K = Deposition velocity [number of particles depositing/(cm²sec)/average number concentration/ cm³ of air].

L = Distance from upstream concentration, C_{o1}, to downstream concentration, C_{o2}, in deposition tube.

P = $P_R + P_S$.

P_R = Number of particles in wind tunnel by-passing the sample probe inlet during the sampling time.

P_S = Number of particles entering the sample probe inlet during sampling time.

Q = $Q_R + Q_S$.

Q_R = Air volume flow in tunnel by-passing sample probe.

Q_S = Air volume flow entering sample probe.

Re = Reynolds number, $DV\rho/\mu$.

S_S = Particle stop distance $\rho d^2 V/18\mu$.

Stk = S_S/D

V = Bulk average axial air velocity in deposition tube.

μ = Air viscosity.

ρ = Particle density.

References

1. Soo, S. L.: *Fluid Dynamics of Multiphase Systems.* p. 165, Blaisdell Publishing Co., Waltham, Mass. (1967).

2. Lundgren, D., and S. Calvert: Aerosol Sampling with a Side Port Probe. *Amer. Ind. Hyg. Assoc. J. 28:* 208 (May-June 1967).

3. Sehmel, G. A.: Estimation of Airstream Concentrations of Particulate from Subisokinetically Obtained Filter Samples. *Amer. Ind. Hyg. Assoc. J. 28:* 243 (May-June 1967).

4. Vitols, V.: Theoretical Limits of Errors Due to Anisokinetic Sampling of Particulate Matter. *J. Air Pollution Control Assoc. 16:* 79 (Feb. 1966).

5. Badzioch, S.: Collection of Gas-borne Dust Particles by Means of an Aspirated Sampling Nozzle. *Brit. J. Appl. Phys. 10:* 26 (Jan. 1959).

6. Hemeon, W. C. L., and C. F. Haines, Jr.: The Magnitude of Errors in Stack Dust Sampling. *Air Repair 4:* 159 (Nov. 1954).

7. Watson, H. H.: Errors Due to Anisokinetic Sampling of Aerosols. *Ind. Hyg. Quart. 15:* 21 (March 1954).

8. Schwendiman, L. C., A. K. Postma, and L. F. Coleman: A Spinning Disc Aerosol Generator. *Health Phys. 10:* 947 (Dec. 1964).

9. Friedlander, S. K., and H. F. Johnstone: Deposition of Suspended Particles from Turbulent Gas Streams. *Ind. Eng. Chem. 49:* 1151 (July 1957).

10. Owen, P. R.: Dust Deposition from a Turbulent Airstream: *Int. J. Air-Water Pollution 3:* 8 (Oct. 1960).

11. Davies, C. N.: Deposition of Aerosols from Turbulent Flow through Pipes. *Proc. R. Soc. A. London 289:* 235 (Jan. 1966).

12. Wells, A. C., and A. C. Chamberlain: Transport of Small Particles to Vertical Surfaces. *Brit. J. Appl. Phys. 18:* 1793 (Dec. 1967).

13. Stavropolous, N.: *Deposition of Particles from Turbulent Gas Streams.* M. S. Thesis, Columbia University, New York (1957).

14. Postma, A. K., and L. C. Schwendiman: *Studies in Micromeritics—Particle Deposition in Conduits As a Source of Error in Aerosol Sampling.* HW-65308, Hanford Laboratories, General Electric Company, Richland, Washington (May 1960).

15. Sehmel, G. A., and L. C. Schwendiman: *The Turbulent Transport and Deposition of Particles Within Vertical Conduits,* HW-SA-3183, Hanford Laboratories, General Electric Co., Richland, Washington (Oct. 1963).

16. Sehmel, G. A.: *Aerosol Deposition from Turbulent Airstreams in Vertical Conduits,* BNWL-578, Battelle Memorial Institute, Pacific Northwest Laboratory, Richland, Washington (March 1968).

17. Sehmel, G. A.: Particle Deposition from Turbulent Air Flow. *J. Geophysical Research 75:* 1766 (March 1970).

18. Sehmel, G. A.: The Density of Uranine Particles Produced by a Spinning Disc Aerosol Generator. *Amer. Ind. Hyg. Assoc. J. 28:* 491 (Sept.-Oct. 1967).

19. *Perry's Chemical Engineers Handbook.* 4th Ed., McGraw-Hill Book Co., Inc., New York, New York (1963).

20. Schlichting, H.: *Boundary Layer Theory.* 4th Ed., p. 530, McGraw-Hill Book Co., Inc., New York, New York (1960).

21. Schwendiman, L. C., G. A. Sehmel, and A. K. Postma: *Radioactive Particle Retention In Aerosol Transport Systems,* HW-SA-3210, Hanford Laboratories, General Electric Co., Richland, Washington (Oct. 1963).

22. Epstein, L. F., and T. F. Evans: *Deposition of Matter from a Flowing System,* GEAP-4140, General Electric Co., San Jose, California (Dec. 1962).

23. Green, H. L., and W. R. Lane: *Particulate Clouds: Dusts, Smokes, and Mists.* 2nd Ed., p. 271, D. Van Nostrand Co., Inc., Princeton, New Jersey (1964).

The Validity of Gravimetric Measurements of Respirable Coal Mine Dust

K. M. MORSE, H. E. BUMSTED and W. C. JANES

While the gravimetric method of assessment of coal dust exposures is superior to the microscopic count method in the ease, speed, and precision of the final measurement, there are shortcomings in the validity of sampling and interpretation. Data and experience with the gravimetric method are presented to demonstrate the significance of these shortcomings in relation to the legal requirements and standards for use of this method in coal mines. Errors inherent in the sampling or introduced by inadequately trained technicians using the method can have unwarranted serious import on coal mine operations. The gravimetric method has not been sufficiently correlated against pulmonary response to inhaled dust in this country to validate a standard to be strictly enforced with possibilities of severe penalties.

Introduction

THE COAL MINE HEALTH and Safety Act of 1969 not only requires coal mine operators to meet the most stringent dust standards in the world, but also requires the application of an instrument for dust sampling which has had very limited use in this country. Therefore, it is pertinent to discuss some of the major problems relating to the validity of measurements of dust concentrations by gravimetric dust samplers. These are of special concern when regulatory agencies are utilizing these instruments to enforce dust standards as exact delineations between safe and hazardous dust concentrations.

It is not the objective of this paper to discount the gravimetric method of dust assessment. It is a simpler method when utilized by qualified persons than the tedious particle-count method, which is time-consuming and thereby uneconomical in the utilization of professional manpower. However, as employed in this country, the accuracy and reproducibility of results of the gravimetric method depend upon the resolution of the inherent errors which arise in the application of any new method. These errors are due to human and instrumental factors which develop among those individuals who have had little experience in the field application of the gravimetric method, or little expertise in the evaluation of environmental factors. Unfortunately, this group comprises the majority of persons who are conducting dust surveys for governmental agencies and the coal mining industry. Additionally, there are some basic unresolved problems on the correct sampling rate and the weighing technique.

Basis for Gravimetric Assessment

The 1959 International Conference on Pneumoconiosis, held in Johannesburg, South Africa, adopted several recommendations for assessing a dust concentration in terms of its pneumoconiosis risk. These were to the effect that (a) dust measurements should be related to the respirable fraction of the dust, (b) the best single descriptive parameter of the respirable coal dust concentration was its mass and (c) the sampling device for collecting the respirable dust should have an elutriator meeting a stated collection performance. These recommendations were adopted by the British Medical Research Council which had been undertaking extensive dust surveys in coal mines together with the National Coal Board. These surveys were a phase of a thorough epidemiological study which was endeavoring to correlate respirable dust concentrations, as evaluated by the particle count parameter, with the progression of pneumoconiosis in a select group of coal miners. It is significant to note that this select group of coal miners were men previously working in the same mines in which their dust exposures were not indicated. After accumulating data on dust concentrations for a period of about ten years, the British concluded that a satisfactory correlation between pneumoconiosis and dust exposure, as measured by the particle count parameter, could not be established. Consequently, they set about to develop a dust sampling instrument which would describe the respirable dust concentration in terms of the mass parameter. This research led to the development in 1964 of the instrument[1] now commonly termed the "MRE Gravimetric Dust Sampler," which is now the reference instrument specified in the standards in the Coal Mine Health and Safety Act of 1969. The British researchers employed this sampler with their previous standard dust sampler (thermal precipitator) to develop conversion factors to enable conversion of previously assembled dust concentration data, described by the particle count parameter, to the mass parameter. This led to a statistical correlation between mass concentration of dust and pneumoconiosis progression. However, the conversion of previous assembled particle count data into mass units is only an approximation due to the varying composition and density of coal mine dust particles.

American investigators were also developing gravimetric sampling instruments, among which was the development of Hyatt.[2] These investigators also developed a two-stage sampling instrument but utilized a small cyclone instead of a horizontal elutriator to separate out the non-respirable and respirable fractions of the dust. As in the case of the MRE sampler, these size-selective devices operate upon the basis of the "aerodynamic diameter" to describe particle size. It is of interest to the subject of gravimetric sampling that both the MRE and cyclone-type of size-selective air samplers were stated to be basically developed upon the pulmonary deposition data of Brown and Hatch.[3] Yet, both instruments were designed with different definitions of respirable dust, as will be discussed later.

Factors Influencing Valid Measurements

When the MRE instruments became available in this country, the authors undertook to test this instrument in coal mines by simultaneous sampling with personal samplers and the midget impinger. Subsequent to these samples, over 2000 gravimetric samples have been collected with the personal sampler, which is approved for dust sampling under the federal coal mine law. The objective of this work was to determine what correlation could be obtained between these instruments and to determine what problems were involved in their field use and in weighing of the filters. The simultaneous sampling was done by mounting a package containing five instruments on the continuous mining machine beside the operator, and also equipping the operator with a personal sampler. Our instrument package contained the following:

1—AEC gravimetric sampler @ 1.4 lpm
1—AEC gravimetric sampler @ 2.0 lpm
1—USBM Midget impinger @ 0.1 cfm
1—U.S. Steel midget impinger @ 0.1 cfm
1—MRE gravimetric sampler @ 2.5 lpm

This package is shown in Figure 1. We have obtained 48 sets of tests with this package. Regression studies of this data were made by our statisticians. The results of these studies indicated a good correlation, based on partial correlation coefficients and RSQ, between the MRE @ 2.5 lpm, the personal sampler @ 2.0 lpm and the impinger.

FIGURE 1.

The views expressed in this paper are based on these above mentioned samples collected in the course of general mine dust surveys, our laboratory analyses, our testing of various types of filter papers, and general observations made in the use of these gravimetric instruments.

The operators of coal mines are given the choice of collecting the required respirable sample by a sampler worn by a cutting machine operator, or by placing the sampler on the mining equipment. The data collected during our study indicated there was no correlation between the personal sampler (@ 2.0 lpm) in the package and the personal sampler (@ 2.0 lpm) worn by the continuous miner operator. Statistical analysis of the data indicated that each sampler could have been sampling a different dust cloud.

We would group the problems in gravimetric sampling in the following categories:

(1) Training problem.
(2) Flow rate disagreement.
(3) Unattended operation.
(4) Pre-weighted filter control.
(5) Filter selection.
(6) Rock dust contribution.
(7) Moisture effects.
(8) Cyclone orientation.
(9) Weighing problems—small mass.
(10) Quartz analysis.
(11) Lack of correlation with pulmonary response.

Training

One of the major problems which continues to be a deterrent to valid dust samples, is the large number of relatively untrained people who are undertaking dust sampling. Obtaining valid dust measurements is not as simple as some proponents of gravimetric sampling have endeavored to project. People who never before were engaged in dust sampling, or who do not have a health conservation philosophy, are now engaged in this work and are considered competent merely by taking a three-day course conducted by the U.S. Bureau of Mines. This course has not provided the training necessary to develop persons competent in dust sampling. Some of the chief inadequacies of the course are that persons are not involved in actual collection of samples and weighing of filters, nor are they provided an adequate understanding of the limitations of the sampling equipment. The instruction is mainly of the didactic type with no field or significant laboratory work. This situation arose due to the lack of time for training afforded by Congress to the Bureau, between the passage of the Act and the effective date of the dust provisions. Six months was grossly inadequate, particularly when a new sampling technique was involved; the use of a pre-weighed filter was a new sampling concept and instrument changes were made by the Department of Health, Education and Welfare. This situation was aggravated by the unavailability of the sampling instruments and the pre-weighed filters until after the effective date of the dust standards. Therefore, there was no opportunity to gain experience with the dust sampling equipment by those who were responsible for administering and conducting the sampling program.

This training problem not only existed in the coal industry but also in the enforcement agency. Mine inspectors, who had no experience in dust sampling for health hazard evaluation, with the exception of a short training course within the agency, are now engaged in enforcing a dust standard, the violation of which can result in severe penalties.

One of the proponents of gravimetric sampling is the Aerosol Technology Committee of the American Industrial Hygiene Committee which is largely composed of governmental and university industrial hygienists. This Committee has developed an *Interim Guide For Respirable Mass Sampling*. In the introduction to this guide is the following statement: "Although several commercial instruments are available for sampling of this type, there has been some disagreement as to their proper operating flow rate. The sensitive techniques required for measuring small mass concentrations and different definitions of respirable dust raise further questions for the practicing industrial hygienist." This statement causes one to be concerned with injecting into this "state of the art" a large number of people who are not industrial hygienists and have little training.

Unattended Operation

An extremely and unnecessarily large number of samples will be required of each coal operator. For example, based on the sampling requirements stated in the Federal Register, a 13 section mine employing 550 miners, would be required to obtain a minimum of about 2200 air samples annually, provided the mine was always in compliance. This number would greatly increase if two or more sections are not in compliance. In order to obtain this large number of samples in the widely distributed working sections of a mine, it is necessary to equip each miner with a sampling instrument just prior to his entering the mine, and retrieve it as he exits the mine. The section foreman must be relied upon to check the flow rate several times during the shift. This results in the sampling instrument being largely unattended and therefore, will be subject to incorrect flow rates or in some cases to tampering by a few individuals who for one reason or another will deliberately contaminate the sample. This can be easily done by creating dust from sources like the floor, clothing, and loose rock. The majority of tampering that we have experienced has been with the sampling head (filter and cyclone). There is no way at present to preclude this and the law does not provide any penalty for those few individuals who may practice it. Therefore, a considerable training and educational program will be necessary to impress upon mine workers that their interests will not be served by tampering. There is nothing in the specification which requires these two parts to be tamper-proof in regard to deliberate contamination of a sample.

Invalid samples can develop from other causes. These can include low air flow due to a pinched hose. This situation can occur when a miner is carrying material like heavy timbers, boards, or roof bolts. When a miner wears the pump on the front of his belt, one specific type can be turned off, when he bends over, by his body depressing the switch.

Pre-weighed Filter

For the first time in the history of air sampling, a commercially pre-weighed filter is required, which is encapsulated and in turn enclosed in a casette. An accurate tare weight on the filter is extremely important if the true dust level is to be determined. In our use of the vinyl-metricel (VM-1) filter, the tare weights varied widely. In one box of one lot number we have found a variance of from 10.64 mg to 17.62 mg. From box to box the variation may be much greater. After weighing several thousand of these filters, we have found weights as low as 9 mg and as high as 22 mg. This would indicate a great variation in the thickness and/or density of the commercially prepared sheets of the filter material and indicates a probable variation in sampling efficiency.

With this wide variation in filter weight, the supplier of the pre-weighed filters will have to weigh each individual filtering unit. Weighing of these units must be accurate to at least 0.1 mg. The production weighing of these units is done manually and is time-consuming. This probably increases the cost of the units. No information has been forthcoming on how the accuracy of the weight of each filter is assured. We conducted a limited study of the reproducibility of the manufacturer's weights on some of the first available pre-weighed filters. Twenty filters were randomly selected from a lot of a hundred. We found that 20% of the filters did not meet the ±0.1 mg tolerance of the HEW specifications and 10% exceeded ±0.2 mg.

137

Failure to do so can result in a dust violation when the fault can lie with inaccuracies in the tare weights of the filters.

Users of these filter casettes are in a sense being required to accept the accuracy of sampling equipment without any opportunity to undertake their own study or review any quality control procedure. Industrial hygienists have repeatedly found major errors in the sensitivity and accuracy of commercial industrial hygiene sampling devices. One cannot accept the view of the Departments of Interior and HEW especially when their experience is no greater than that of experienced industrial hygienists in industry. It is basic in any industrial hygienist's philosophy to question the accuracy of any sampling device until he has undertaken to calibrate it.

Filter Selection

The personal sampler specifications of HEW state that the filter shall be a membrane filter type with a nominal pore size of not more than 5 microns. This infers that smaller pore sizes can be used. In our work with 37 mm membrane filters, we have found considerable variation in the pressure drop across such filters of different pore sizes under 5 microns. This may be reduced by increasing the filter area for the required flow rate. In some very limited work in simultaneous sampling with several filters of the same pore size and with Whatman-41 (W-41) filters, we have found appreciable differences in the amount of respirable dust collected on each filter. The weights of dust on the W-41 and VM-1 filters were quite similar, with appreciably lower results on the Millipore and Nuclepore filters. The low results obtained with the Nuclepore filters were expected since photomicrographs have shown this filter to have clearly defined holes of a stated pore size through the filtering material. The VM-1 and Millipore filters appear more sponge-like. The filter specifications state that the filter shall be unaffected by isopropyl alcohol. The VM-1 filter shows appreciable weight loss in

138

isopropyl alcohol while the Millipore filter is severely affected.

We have reservations about the use of VM-1 filters for several reasons. We do not prefer it if we are to undertake free silica analysis by chemical procedures. We have found that the 37 mm membrane filters have a very high ash content from 19.35% to 23.42% of the tare weight. Examination of this ash by x-ray diffraction has shown it to contain appreciable amounts of alpha cristobalite. Analyses of many different lot numbers of filters have shown that the free silica content will vary from 0.2 to 1.8% of the filter weight. The amount will vary from box to box and even from filter to filter in the same box. Analyses of three filters from each of three different lot numbers showed the free silica content to vary from 0.29% to 1.61% of the filter weight. The free silica content has been checked by the Talvitie, Jephcott and Wynne-Trostel chemical procedures as well as by x-ray diffraction. Because of the wide variation in free silica content from filter to filter, it is not possible to apply an average correction value.

To illustrate the resulting problem of this filter blank, assume a filter containing 1.0 mg of coal dust with a 2% free silica content. The free silica in the coal dust would be 0.020 mg. If the filter weighted 15.00 mg and contained 0.3% free silica, its contribution upon ashing would be 0.045 mg. Thus, the ash of the filter and dust samples would contain 0.065 mg of silica, which would be 6.5% of the weight of the coal dust. The dust would show a percentage of free silica of more than three times the true value. We have also determined the ash content of a number of other filters, such as W-41, Millipore, GA-1, Nuclepore and Satorius and all were under 0.07% of the tare weight with the Nuclepore and Satorius having practically no ash.

In our tests on filtering efficiency, we have found the W-41 to give an efficiency equal to the VM-1 and, as a matter of fact, frequently

139

use this filter. While W-41 filters are hygroscopic, we have been able to use this filter by keeping the new filters in a dessicator cabinet, weighing on a balance with dessicator cylinders in the balance cabinet and after use, storing in the dessicator cabinet for at least 16 hours before weighing. This procedure has proven very satisfactory over the years. It is actually our practice to follow this procedure with all types of filters. Furthermore, we have had significant moisture build-up on the VM-1 filter which is considered as a non-hygroscopic filter.

Rock Dust Problem

The new law requires rock dusting with lime stone to within 40 feet of the working face. Since this limestone dust is quite fine, it is readily dispersed into the air by equipment movement and the miner's movements. Unfortunately, this generated non-coal dust is generally carried to the working face by the ventilation being supplied to the face for dust control. Therefore, the federal coal mine law requires the addition of dust for safety reasons but provides no correction factor for this dust in the air samples required. In other words, all the collected airborne dust is considered as coal dust. As the limestone is very fine once it becomes airborne, it will be largely of the respirable size.

Recently we have developed an x-ray technique which will permit the estimation of the rock dust content of the individual air samples. In two mines we have found as high as 50% of the sample to be calcite or calcium carbonate. Many of the settled dust samples picked up in these mines at the continuous mining machine were gray in color. While no attempt was made to determine the exact amount of calcite present from the x-ray diffraction pattern, it was estimated that more than 30% of the settled dust was calcite.

Moisture Effects

The specifications for dust sampling instruments state that the filter shall be non-hygroscopic and membrane filters have been placed in this category. We have loaded 37 mm VM-1 filters with coal dust, arranged them in series with a fritted bubbler containing water and a chamber for removal of water droplets. We drew air through the chain at 2.0 liters per minute and weighed the filters after 30, 60, 90, and 120 minute intervals, returning the filter to the chain after the first three intervals. After 120 minutes, the first filter gained 16.52 mg of moisture with the subsequent filters in the chain showing lesser but appreciable moisture buildup. By placing the filters in a dessicator for 25-30 minutes, the filter weights will generally return to the dry weight. We have similarly added moisture to non-dusted VM-1 filters. Even if one assumes a filter is non-hygroscopic, this does not mean that coal dust and rock dust are not hygroscopic.

In a recent coal mine study using VM-1 filters, 110 filter holders were sealed tightly after sampling. When they were returned to the laboratory, the filters were weighed immediately after opening the holder and again after 48 hours in the dessicator. Loss in moisture varied from 0.00 to 0.68 mg on drying where the dust loadings varied from 0.03 to 2.62 mg. In casual examination of the data, there appeared to be no relationship between either filter weight or dust loading and the amount of moisture. While the average moisture pick-up was relatively small and would appear to be of little significance, what concerns us is that unless the moisture effect of these samples were taken into consideration, we would have had five samples which would have indicated non-compliance with the standard. When these standards are to be enforced by mine inspectors weighing on the site, any method which can result in about 5% of the samples indicating non-compliance due to moisture effect should be modified, especially when the test data can result in a citation or a closure order. However, no procedure has been announced as to how the dust samples are to be handled. Once again, we are expected to accept, without qualification, the experience of governmental agencies who have as little, if not less, experience in the application of the gravimetric technique as some non-government hygienists who have been using it for several years.

Cyclone Orientation

A cyclone is said not to be sensitive to position orientation as far as the theory of operation is concerned. However, we believe that a 10-mm cyclone is affected as an elutriator by its position. In low coal, when it is worn by repairmen, or during handling bulky and heavy materials, the device does not always hang in a vertical position due to the position of the wearer, or movement of the tubing between the sampling head and pump. Under such conditions, it is possible to pull some of the non-respirable dust from the cyclone into the filter which collects the respirable dust. In extreme cases some of the contents of the cyclone have been dumped onto the filter. We and others have found this occurring from time to time with the result that the respirable dust concentration is highly elevated and is obviously incorrect. If one of these samples occurs in the sample cycle, the average or total concentration for the entire sample cycle would exceed the dust standard, regardless of the fact that all the other samples were low. The Bureau of Mines has indicated that all dust samples above 6.0 mg/m^3 will be examined optically to determine if oversized particles are present. Such oversized particles can be collected in any sample. Therefore, any sample exceeding the standard should be so examined. In the case of an over-anxious inspector, a mine could be improperly cited for a dust violation.

It has been our experience that the greatest possibility for a cyclone to be inverted, resulting in dumping non-respirable dust into the filter, is when the unit (filter head and pump) is being removed from the worker at the end of the shift. This usually occurs when the worker tries to remove the unit himself. Almost invariably he detaches the filtering head first and then allows it to drop while using two hands to remove the pump. As mentioned above, this can lead to a highly elevated dust sample. This again stresses the need for training and education of the mine workers.

Weighing of Small Mass

The extension of weighing techniques from macro to the micro range creates many problems. With the present semi-micro balances under the best conditions, weighings by an experienced operator will vary by about $\pm$ 0.03 mg. These balances must be kept free from vibration, air currents, and temperature changes. The zero setting on the balances must be checked frequently to eliminate the zero drift and the linearity should be checked periodically. In one make of semi-micro balances the internal weights are of a Class M specification which requires a $\pm$ 0.10 mg tolerance on the weights. In this instance there could be differences of several tenths of a milligram between two different balances in weighing the same 5 gram weight.

Our laboratory has, at present, two semi-micro balances of the same make, one four years old (A) and one new (B). In checking weights on the two balances, A was 0.11 mg higher than B for a 12.50 mg filter. However, on a 5 gram weight, B was 0.36 mg higher than A. This illustrates the potential problems which can develop by the use of two different balances.

It has been the practice in our laboratory over the years to weigh the used filters on the same balance used for determining the tare weights. If at all practical, the same individual will weigh the filters both times. This reduces the errors due to the idiosyncracies of both the balance and the operator.

The proposed procedure of the manufacturer of the filter weighing the unit for tare values and the Pittsburgh laboratory of the U. S. Bureau of Mines or the mine inspector weighing the loaded filters present opportunities for errors. Semi-micro balances are not generally classified as portable and unless the mine inspector is adequately trained in the care and maintenance of the balance, the accuracy could deteriorate rapidly. It would be anticipated that vibration, temperature and air currents could create severe problems in weighing at a mine office or motel room.

In weighing, it is considered to be good practice to keep tare weights as near as possible to the amount to be determined. The accurate determination of 1 or 2 mg dust loading on a filter unit weighing 5000 mg is a very difficult task.

With all the above real opportunities for inaccurate weighing, one should question with good conscience citations for violations where the dust standard has been exceeded by 0.3-0.4 milligrams.

Quartz Analysis

The classical x-ray diffraction and chemical methods for the determination of free silica are not sensitive enough to be used for individual air samples of 1 to 2 milligrams. In x-ray diffraction techniques, the best accuracy is gained by the use of an internal standard to correct for matrix effects. In air samples, this is not possible.

Both the Talvitie and the Jephcott procedures have been studied extensively in our laboratory. They have not proven to be of value due to the solubility in phosphoric acid of the extremely small quartz particles found in respirable dust samples. These methods were originally developed for the macro analysis of settled dust samples. We have not been able to adapt them for the small samples of 1 to 5 micron dust. Since for these techniques it is necessary to ash the filters, the free silica blank in the filter becomes a major problem. While the final colorimetric estimation of the silica content has sufficient sensitivity for individual air samples, we have not been able to solve the problems associated with the phosphoric acid separation of the silicate minerals.

At present the infrared technique does not have the sensitivity to determine the free silica content of the individual air samples. The Bureau of Mines technique involves the partial removal of the dust from the filter in an ultrasonic bath and the combination of several samples to get sufficient material for analysis. The removed dust is ashed in a micro-furnace. A weighed portion of the ash is then pressed in a KBr pellet and the silica content determined by the infrared absorption at 775 and 796 wave numbers. The minimum amount readily detectable is about 40 micrograms.

There are several inherent problems with this procedure. First, there is no published data to show that the dust removed from the filter is a representative sample of the dust on the filter. It is possible that the lighter larger particles of coal dust might be more easily removed from the filter than would the heavier mineral dust. We have not been able to prevent filter degradation in the ultrasonic bath. Clean VM-1 filters have been treated for one minute in an ultrasonic bath with isopropyl alcohol and water as the solvents. Losses of 1.27% in water and 3.66% in alcohol were found based on the initial filter weight. This loss appears to be of a particulate nature rather than solution of the filter material. This released filter material contaminates the dust sample removed from the filter. In the case of alcohol washing as few as five filters can add 2.5 mg of filter material to the dust sample. This added material can greatly affect the free silica determination.

In the infrared analysis of particulate material, the absorbance is affected by particle size. Thus, unless the calibration curve is prepared with quartz of the same particle size range as that of the material to be analyzed, the calibration curve may give erroneous results. Nevertheless, infrared techniques are being used in England and Germany and should be carefully studied to determine whether this analytical procedure can be applied to individual respirable dust samples.

Recently, we have been able to extend the detection level of our x-ray equipment to

0.020 mg of quartz on a VM-1 filter. Fine silica dust was loaded on VM-1 filters through a standard 2-stage sampling unit operating at 2 lpm. The amount of quartz on the filter was determined by careful weighing. The loaded filters were placed in a specially designed holder and the 3.35°A line of quartz was scanned using a scintillation counter and a carefully aligned pulse height discriminator. A plot of peak height versus milligrams of quartz was prepared and was found to be linear up to 1 mg of quartz.

A similar procedure was carried out using calcite and measuring the peak height of the 3.04°A line. Again, the curve was linear to 3.0 mg calcite.

In this manner it was possible to estimate the quartz and calcite on individual filters where the dust loading was 0.20 mg. It should be pointed out that both the infrared and the x-ray diffraction techniques do not determine other forms of free silica.

When a technique for the removal of the dust from a filter without contamination with filter residue is developed, an internal standard can be added to the removed dust and a more accurate estimate of the silica content can be made by x-ray diffraction.

Further study toward the refinement of both the x-ray diffraction and infrared methods should be undertaken. The recent development of the high intensity x-ray tubes may in time greatly improve the sensitivity of this method. However, both procedures require large investment in equipment which may not be practical for the small mine operator. For these the development of a simple chemical procedure is of prime importance. Nevertheless, it appears that infrared techniques which are widely used in England and Germany should be intensely examined for this analytical procedure.

Need for Medical Validation

In our opinion, too often the industrial hygienist endeavors to by-pass the physician in establishing the validity of a dose-response relationship and equally so, the physician bypasses the industrial hygienist. It should be fully appreciated that occupational health problems will not be adequately resolved unless the physician and industrial hygienist work as a team in undertaking epidemiological studies of a dose-response relationship to establish valid standards.

The valid establishment in this country of a dust standard using a new parameter, such as mass concentration, can be done only by a prospective epidemiologic study of the coal mine environment, together with the medical findings derived from periodic physical examinations. No change in the parameter for describing a dust concentration from one established epidemiologically can be validated alone by correlations of the parameters of different sampling instruments. Therefore, while a gravimetric dust standard has been established in England, the same method for validation must be done in the United States. The use of the standards from the English study and merely reducing them for the sake of having a lower number can be termed "playing the numbers game" and not in the interests of either the industrial hygiene or industrial medical professions.

Flow Rate Disagreement

Since the availability of the MRE instrument and the American two-stage cyclone type personal sampler, there has been considerable discussion on the different definitions of respirable dust which these two instruments provide. As previously mentioned, the MRE elutriator performs with an elutriator performance curve developed by Davies and approved by the British Medical Research Council. The cyclone personal sampler performs on a separation curve approved by an AEC Committee meeting at Los Alamos Laboratory. The difference in the performance in these two elutriators and consequently, the definition of respirable dust is shown in Figure 2. As can be observed, the personal sampler collects 50% of all particles

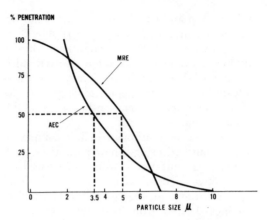

FIGURE 2.

equivalent to a 3.5 micron spherical particle of unit density. The MRE collects 50% of all particles having a falling velocity equivalent to a 5-micron unit density sphere.

It should be noted that both instruments were originally designed and calibrated on unit density spherical particles and both collect other particles supposedly in relation to their mass equivalents to a unit density particle. This concept may be better understood by reference to Table I which shows equivalent particle diameters for water, coal and quartz. By application of equivalent diameter concept, and in reference to the MRE elutriator performance curve showing 50% collection of 5.0 micron particles of unit density, this instrument would collect 50% of all quartz particles having a mass equivalent to a 3.1 micron spherical quartz particle or a 4.5 micron spherical coal particle. It is to be noted that about all the early work on calibration of personal samplers was with unit

TABLE I

Equivalent Particle Diameters for Various Densities Compared with Unit Density

Material	Water	Quartz	Coal*
Density	1	2.6	1.3
Particle Diameter (Microns)	0.5	0.31	0.45
	1.0	0.62	0.89
	2.0	1.24	1.78
	3.0	1.86	2.67
	4.0	2.48	3.55
	5.0	3.10	4.50
	7.0	4.34	6.28
	10.0	6.20	8.90

*Pure coal.

density spherical particles. However, industrial dusts, particularly coal dust, are anything but spherical particles. As the shape of the particle has a significant effect upon its falling speed, there would be some doubt about calibrating the first stage (elutriator) of these gravimetric samplers against spherical particles of unit density and calculating its performance against irregular shaped particles by the equivalent diameter concept, which does not adequately take into consideration the effect of the shape of the particles. The effect of shape and the different densities of the other components upon coal mine dust is demonstrated by Hamilton and Walton[4] of the National Coal Board of England is shown in Figure 3. Obviously, there is a considerable effect which demonstrates why one cannot make a correlation between two different instruments, which describe the dust concentrations by different parameters, and change a dust standard based upon one parameter to that of the other without studying the prevalence and progression of pneumoconiosis. This is the method employed by the Medical Division of the National Coal Board in deriving their new gravimetric standards.

American and Canadian investigators have also devoted considerable effort to the determination of the correct flow rate of the 10-mm cyclone used in personal samplers. Monodispersed and polydispersed aerosols have been used in these efforts to determine the

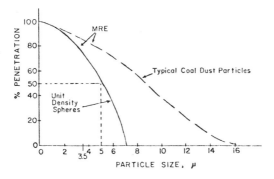

FIGURE 3.

149

flow rate that produces the separation characteristics of the cyclone which conforms to the definition of respirable dust of the AEC-Los Alamos meeting. This has resulted in different investigators recommending flow rates of 1.4, 1.5, 2.0, 2.5 and 2.8 liters per minute. However, Tomb[5] has done this using an aerosol composed of coal dust and he has shown that a flow rate of 2.0 liters per minute conforms to the AEC recommendation for the performance of a two-stage personal sampler of the cyclone type. In our view, calibrations to meet the performance of an elutriator, which in turn is supposed to meet a certain definition of respirable dust are only a part of the requirements of a good instrument. What is more essential is that the instrument performance have reproducibility. However, its use in changing the parameter of a dust standard must be based upon its ability to be correlated with some index of the disease incidence or progression, similarly as done by the National Coal Board. If this is demonstrated with any instrument that exhibits reproducibility, the instrument parameter is of secondary importance. Such biological correlation applies only to the dust under study and similar correlation must be done for other industrial dusts. This has not been done for coal dust in this country although such is underway.

At the sampling rate of 2.0 liters per minute, the U. S. Bureau of Mines has stated several formulae for the linear relationship between the MRE and the 10-mm cyclone sampler. The latest of which we are familiar was the following: $MRE + 0.6 = 1.6 \text{ P.S.}$, where P.S. is the dust concentration measured with the personal sampler. Based upon our 48 sets of test data, using the 2.0 lpm sampling rate, our statisticians expressed this relationship as $MRE = 0.8 + 1.4 \text{ P.S.}$, and this expression gave results very close to the Bureau of Mines expression. However, the Interim Compliance Board under the Coal Mine Health and Safety Act published the official relationship between the two instruments as, $MRE = 1.6 \text{ P.S.}$ Therefore, this states that

3.0 mg/m³ measured on the MRE is equivalent to 1.9 mg/m³ measured by the personal sampler.

Conclusions

We have attempted to point out several problems involved in the application of gravimetric measurements of respirable coal mine dust, and its use for establishing dust standards. There are advantages in this type of instrument over particle count instruments involving microscopic counting. However, the parameter determined by gravimetric dust samplers has not been correlated against pulmonary response to inhaled dust. Until this is done in this country, a gravimetric dust standard is premature and more a "guesstimate" and not a particularly good one, than a valid standard which is to be enforced with severe penalties for violation.

References

1. DUNMORE, J. H., R. J. HAMILTON, and S. G. SMITH: An Instrument for The Sampling of Respirable Dust for Subsequent Gravimetric Assessment. *J. Sci. Instr.* 41: 669 (Nov. 1964).
2. HYATT, E. C., H. F. SCHULTE, C. R. JENSEN, R. H. MITCHELL, and G. H. FERRAN: A Study of Two Stage Air Samplers Designed to Simulate the Upper and Lower Respiratory Tract. *Proceedings 13th Intl. Cong. on Occup. Health,* New York (1961).
3. BROWN, J. H., K. M. COOK, F. G. NEY, and T. F. HATCH: Influence of Particle Size Upon the Retention of Particulate Matter in the Human Lung. *Amer. J. Public Health* 40: 450 (1950).
4. HAMILTON, R. J., and W. H. WALTON: *The Selective Sampling of Respirable Dust; Inhaled Particles and Vapours,* Pergamon Press, London (1961).
5. TOMB, T. F., and L. D. RAYMOND: *Evaluation of the Penetration Characteristics of a Horizontal Plate Elutriator and a 10 mm Nylon Cyclone Elutriator.* Report of Investigation—7367, U. S. Bureau of Mines, Pittsburgh, Pa. (March 1970).

AUTHOR INDEX

KEY-WORD TITLE INDEX

152